ENGINEERING
DESIGN

ENGINEERING
DESIGN

A SYSTEMATIC APPROACH

By

DR.-ING. ROBERT MATOUSEK

MUNICH

Translated from the German by

A. H. BURTON

and edited for British readers by

PROFESSOR D. C. JOHNSON

M.A., M.I.Mech.E.

Professor of Mechanics, University of Cambridge

INTERSCIENCE PUBLISHERS

A DIVISION OF JOHN WILEY & SONS, INC.

PUBLISHED IN GREAT BRITAIN
BY BLACKIE & SON, LIMITED, LONDON AND GLASGOW
AND IN THE UNITED STATES OF AMERICA
BY INTERSCIENCE PUBLISHERS INC.
NEW YORK

First published
1963
©
BLACKIE & SON LIMITED 1963

The German edition of this book entitled
'Konstruktionslehre des allgemeinen Maschinenbaues'
is published by Springer-Verlag, Berlin
Göttingen, Heidelberg
First published, 1957

PRINTED IN GREAT BRITAIN BY BLACKIE & SON LIMITED · GLASGOW

PREFACE

TO THE GERMAN EDITION

This book is addressed to those engineering students who are prepared to work—not to such as are content to refurbish existing designs without taking the trouble to understand the trains of thought and the considerations which are needed in true design work.

It is a well-established fact that the beginner, confronted by the simplest of design problems, and lacking a pattern or model to suggest a solution, loses his way in endless trial and error unless given positive guidance. In this book, therefore, the author has drawn on his long teaching experience in an attempt to present in a readily understandable and systematic manner a methodical work-plan which will enable the beginner practising design problems to reach his objective by a rational route. This approach has the further advantage, confirmed by experience, that in adopting it the student will find his interest and pleasure in design work growing, and his self-confidence increasing.

Written with the requirements of general mechanical engineering in mind, the book does not deal with the manufacturing methods typical of light precision engineering.

To prevent the book from taking on a size which would have detracted from its clear layout and obscured the main principles presented, the numerical tables, graphs, etc. available for reference in pocket books and textbooks have been omitted.

The author is well aware that no beginner can possibly learn the technique of designing from a book which consists only of a collection of principles and rules. In this field only practice makes perfect, and for this reason the text contains a number of exercises.

The book also includes a bibliography of publications dealing with individual aspects of design problems of a kind specially suitable for the student.

May this book awaken in the young engineer an appreciation of the importance of working to a clear plan, for by so doing and by mastering the technique of designing he will come to experience profound pleasure in the beauty of creative work.

Special thanks are due to the publishers who, in their usual manner, have devoted great care to the production of this book.

R. Matousek

Munich
Autumn 1956

PREFACE
TO THE ENGLISH EDITION

The subject of design in engineering occasions much discussion at the present time. It is said by many that far too few trained engineers in this country wish to devote themselves to it and by some that there is insufficient teaching of design in our academic institutions. The position in Germany is different because there engineering education has, by tradition, a considerable design content. This accounts for the fact that the present book was first published in that country; there is, so far as I know, no comparable English text.

It is hoped that this translation will help students and others here to think more about design. In particular it may be of interest in the Colleges of Advanced Technology, where new forms of engineering education may be expected to evolve in the coming years.

D. C. Johnson

Cambridge
Autumn 1962

CONTENTS

INTRODUCTION

The significance of design

The vast strides made by engineering in the past few decades are due primarily to close cooperation between scientists, designers, and production experts. The designer's role in this activity is to be an intermediary between scientific knowledge and the production side. Since his part often fails to attract for the designer the prestige which is his due, it must be pointed out that he has the very responsible task of satisfying in the best manner possible the conditions laid down in the customer's order, and thereby providing the essential foundation for economic manufacture. The finest workshop facilities with the most up-to-date machine tools enabling economic manufacturing methods to be used are of no avail if the designer has not done his work satisfactorily.

High-grade work on the shop floor is only possible if the design itself is good. Again, even the best of salesmen is powerless if the designer has not approached his task with due regard for economic factors and kept down manufacturing costs to render them competitive. The work done by the designer is therefore of fundamental importance to the industrialist and to the whole economy.

Recognition of this fact has led practical men to refer repeatedly in the literature of the subject to the importance of a fundamental training in the art of design. For the same reason, large industrial concerns began many years ago to compile sets of examples of " good " and " bad " practice to keep their engineers' attention focused on some of the rules to be observed if a design is to be successful from the viewpoint of production, assembly, etc. Many component manufacturers publish guides for the use of designers who wish to utilize their components. This applies, for instance, to rolling bearings, belts and chain drives, and oil seals. In recent years, aspects of the problems concerned with appropriate choice of material and correct design have been discussed again and again in journals and books—proof of the importance attached to successful effort in the field of design.

There are some students of mechanical engineering who say: " Why should I have anything to do with design, after all I'm going into the production or the sales side ". Such students have not yet recognized the advantages to be derived from a study of design problems. It is for this reason that many firms of wide experience insist that newly appointed junior engineers shall first spend some time in the drawing office before passing on to the works or into other departments.

An engineer in the shops who has had design training will approach production work with a quite different understanding and will save himself the trouble cf querying many points with the design office. Is it possible to imagine an engineer who would offer an expert opinion, yet who had no idea of the working principle of the machine in question, or of the operation of its various components, or of the advantages and disadvantages of given design arrangements? It is often necessary for a representative to give information on design details to a customer familiar with technicalities, and indeed, even the drawing office itself will often call for design suggestions from one of the firm's representatives.

For an engineer in an administrative post not the least valuable asset of his drawing office experience is the appreciation that he gains that design is a responsible and intellectually demanding task which cannot be undertaken as if it were merely routine work.

Duly recognizing how important design experience is for all engineers, the Verband Deutscher Elektrotechniker* has issued a memorandum on the training of electrical engineers which contains the following passage:

Design is of the utmost importance in the training of an engineer, no matter in what field of activity he may subsequently be employed. A student who has reached a certain standard of capability in design and has found pleasure in it will find things considerably easier when he starts work, even though the path he takes does not lead to the drawing office. For many top posts this is extremely important. The lack of adequate design capability is a deficiency which can be made good only in exceptional cases. Many outstanding men confirm again and again that they themselves have derived great advantage from having spent several years in the drawing office, and their experience shows that a good course of design practice undertaken as a part of technical training exerts a beneficial effect on an engineer's work at all times, regardless of whether he is employed in the planning department, on production, in the laboratory, or on the management side.

It is therefore easy to understand why, in most advertised vacancies for junior engineers, great importance is attached to thorough training in design.

---

* Equivalent of British Institution of Electrical Engineers.

I. GENERAL ASPECTS OF THE DESIGNER'S WORK

What is meant by design?

An observer watching a designer at work will note that when starting on a new assignment he first of all makes a close study of the conditions to be fulfilled. He then ponders the problem for some time before preparing one or more simple schematic diagrams. Perhaps he will also take up his slide rule to check quickly some of the figures involved before resuming consideration of the various possible solutions. Not until the unit or machine has taken shape in his mind does he decide to make several different properly-scaled views of it by a process of alternate calculation and drawing. While thus engaged, however, he has still to consider which material is most suitable, which manufacturing method is most economical, and how the method finally chosen will affect the design. These, and many other points besides, must all be taken into account. Enough has already been said to show that designing is for the most part a purely intellectual, and indeed creative, activity which, contrary to the popularly held view, cannot be regarded solely as draughtsmanship.

The designer is also often widely referred to as a draughtsman. Draughting, however, denotes only that aspect of designing or planning which is concerned with the actual preparation of drawings. Not until the design has developed into a clear picture as seen by the mind's eye— and every design is formed in the mind to begin with—can it be draughted on paper.

Nor can planning be used as an alternative term for design. Planning is rather the preparation of schemes for the use of land, buildings, and industrial equipment.

It will be seen, therefore, that it is not easy to define design activity in a way which covers all the aspects. One thing is certain—in design the main burden of the creative work done is undoubtedly intellectual in nature, and it is intellectual activity of an extremely complex kind. Viewed from a higher vantage point it certainly includes all design procedures, the pure craft activity of drawing, considerations of various kinds—physical, technological, production engineering, mathematical, and economic—as well as the purely formative activity.

The art of designing can perhaps be explained on the following lines.

3

The designer uses his intellectual ability to apply scientfiic knowledge to the task of creating the drawings which enable an engineering product to be made in a way that not only meets the stipulated conditions but also permits manufacture by the most economic method.

What kinds of design work are there?

As in every field of human activity so also in design work there are different degrees of difficulty.

In practice the kinds usually recognized are adaptive designs, developed designs, and new designs.

Adaptive design

In the great majority of instances the designer's work will be concerned with adaptation of existing designs. There are branches of manufacture where development has practically ceased, so that there is hardly anything left for the designer to do except make minor modifications, usually in the dimensions of the product. Design activity of this kind therefore demands no special knowledge or skill, and the problems presented are easily solved by a designer with ordinary technical training. I have often been asked by engineers why I am not content to allow students to " design " from proven existing models. This question will be considered later, but the principal reason is that such a method completely fails to train the design capabilities of the student engineer.

The man who is accustomed to working entirely from existing designs, and who is therefore sometimes called a " pantograph designer ", will not learn to appreciate what designing means until he is confronted with a task requiring original thought, no matter how simple it may be. Of course every beginner must first prove his worth in the field of adaptive design. Unfortunately many " designers " do not get any further.

A rather higher standard of design ability is called for when it becomes necessary to modify the proven existing designs to bring them into line with a new idea by switching to a new material, for example, or to a different method of manufacture. Examples of this will be given in a later section.

Development design

Considerably more scientific training and design ability are needed for development design. Although here, too, the designer starts from an existing design, the final outcome may differ quite markedly from the initial product.

New design

Only a small number of those engineers who decide on design as a career will bring to their work personal qualities of a sufficiently high order to enable them to venture successfully into new design fields. History has many examples, such as the steam engine, the locomotive, the motor car, the aeroplane, to show how difficult it is to design success-fully without a precedent.

Organization of the drawing office

In practice it has become customary to use different professional titles corresponding to the various levels of design activity.

The adjoining diagram (fig. 1) sets out the organization of the staff responsible for design work in a drawing office.

Technical Director
|
Chief Engineer
|
Department Manager
|
Engineer in Charge
|
Deputy Engineer in Charge
|
Section Leader
|
Detail Designer
|
Assistant Designer
|
Designer Draughtsman

Fig. 1.—Organization of personnel in a drawing office

According to a proposal of the professional institution of German engineers, the term *design engineer* should be applied only to engineers who are engaged on design and who, by virtue of special natural gifts and an excellent knowledge of mathematics, physics, and engineering, are qualified in the best sense to undertake entirely independent work.

These are qualities which are certainly called for in a chief designer (who may also be a director), head of department, chief engineer, and deputy chief engineer, that is to say in those engineers who also have to carry a large measure of responsibility. Heads of sections should also have some ability to work without guidance and the capacity to resolve problems without an existing design to copy.

A detail designer, on the other hand, needs only an ordinary standard of professional training on the lines provided, for example, by a technical institute.

It is not intended to imply, however, that engineers trained in this way are not suited to become design engineers. Indeed, it is a fact confirmed by experience that many who have passed engineering school courses are doing outstanding work as design engineers in highly responsible positions.

The engineering draughtsman and the trainee designer are usually ambitious juniors who have been transferred from the shop to the drawing office for training which is carried out there and by attendance at a technical college.

A point which should not be left unmentioned is that in industry appointments to senior design posts are not made on academic qualifications alone; only knowledge and ability are decisive and are made so by the uncompromising demands of industrial practice. Anyone possessing the enthusiasm and drive to improve his knowledge can certainly advance from the position of draughtsman to that of an independent designer.

Relationship between the drawing office and other departments

The two principal areas of technical creative activity are *design* and *production*. The importance of the creative work of the designer is apparent from the single fact that he is responsible for putting the engineering product into such a form that it can be manufactured in the most economical way. Design and production are therefore seen to be closely interrelated. This fact has led to the practice, now common in many works, of bringing the responsible executives together from time to time for an interchange of experience with the design engineers.

But there are also reciprocal relations between other departments of a factory and the drawing office. Results obtained in the test department can often lead to major improvements, or perhaps the design principles of a new product have still to be worked out. That the closest of relationships must exist between the development engineer—or his department if the works is a large one—and the drawing office goes without saying. From the sales engineer, too, the designer can often hear of vitally important points of view which will influence his work.

Official regulations sometimes play a decisive part in determining design; in this connection one need only think of the regulations for steam boilers as laid down by the authorities responsible for structural safety and fire prevention.

The drawing office must of course maintain the closest contact with the customer to ensure that his requirements are clearly understood and catered for as fully as possible.

The various interrelationships involved are shown in diagrammatic form in fig. 2.

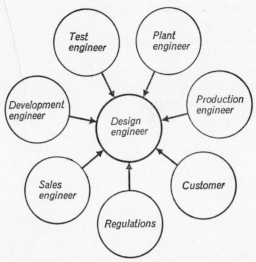

Fig. 2.—Relation between the drawing office and other departments

Why teach design as a special subject?

During his training in the various fields of knowledge the young engineer is crammed with a vast amount of theoretical material and information. He only realizes his helplessness when he is faced with the task of logically applying what he has learned to a specific end. So long as his work is based on familiar models or previous designs, the knowledge he possesses is perfectly adequate to enable him to find a solution on conventional lines. As soon as he is required to develop something already in existence to a more advanced stage, however, or to create something entirely new without a previous design to guide him, he will fail miserably unless he has consolidated his knowledge in depth and so worked upon it that he has reached a higher level of understanding. A detail designer can manage, if he must, with knowledge pure and simple. The design engineer, on the other hand, must have learned to think independently, to reduce logically and to draw conclusions, and to combine. There are many who believe they can acquire all this by attending lectures and reading textbooks. What they fail to realize, however, is that they are only accumulating one fresh item of knowledge after another. Understanding, coupled with powers of logical deduction and judgment, is not a capability that can be conferred from outside; on the contrary, it is something purely personal and inward acquired only by diligent thinking and working with the knowledge already possessed. As mentioned earlier, it is a basic pre-condition for independent designing, and its possession qualifies the designer assisted by a lively imagination to do original work.

In the past, instruction in designing was given by setting the student a problem concerned with the design of prime movers and driven machines. Without any further preparation in design thinking he was then left alone with the problem. The result was that the student looked around for a good existing design, and, having found it, proceeded to work out some leading dimensions; this done, he would start to reproduce the original. The value of such pantograph work as an intellectual exercise was exceedingly small, for the student had no need to rack his brains any further about the construction of the machine or about the kinematic interplay of the various components, or about the form given to the components, or about problems associated with materials and manufacturing methods— the ready-made answer to all these points lay in the original design. The points to be considered in the design had already been worked out, probably by generations of designers in a process of laborious study and painful experience. It is obvious that this method of teaching design only turns the beginner into a copyist, a painter of portraits, because he is ignorant of the entire complex of design thinking.

Even when the beginner is set a problem involving the use of an existing design, he must ask himself the question: Where and how shall I start? This is where the first difficulty makes its appearance. Usually he will begin by looking around for formulae and will discover that he must use his own discretion in employing the rules of mechanics, kinematics, etc. The groping around, and the trial-and-error methods typical of the beginner, are responsible for the view that designing is essentially intuitive. More particularly, it is constantly being emphasized in this connection that designers are born, not made, or in other words that one must be talented for the part. This, however, is an obvious requirement. No one will dispute that all intellectual and manual occupations call for talent. The high qualities required of the designer in this respect in particular, however, are shown below.

Those who are continually pointing out that designing calls for a special talent are giving expression to the view that designing cannot be taught. An analysis of design work carried out from the professional side has led in the last decade or two to recognition of the fact that the technique of designing can also be taught systematically in just the same way as the basic technicalities of all other professions are systematically imparted at schools and colleges.

In connection with the training of designers one often hears it objected, particularly by students who realize their own shortage of talent and would like design to involve no more than copying from an original, that designing is something one can only learn through practical experience. Agreed! Mastery does indeed depend on a great deal of exercise with practical problems. But this applies equally to all other professions. Whoever is content to copy existing designs year after year, however, will never reach the status of the independent designer.

Manufacturing processes have already been perfected to a degree that guarantees optimum output per unit of time in return for minimum outlay. The designer who fails to provide a basis for economic manufacture has not kept abreast of developments in production. The great diversity of the solutions presented in answer to a specific problem involving closely circumscribed conditions is indeed proof that designers are often in some uncertainty about the number of available methods most capable of serving the required purpose, and that in many instances they are still very much in the dark about the extent to which choice of material, product design, and component arrangement can cheapen manufacture, simplify assembly, and promote reliability, etc.

It would be a mistake to overlook the considerable progress that has already been made in educating the young designer in rational methods of working. The same problem forms the subject of many articles in technical journals and books. For the beginner, however, it is difficult to distinguish what ideas are fundamental in such an abundance of published material. Above all there is the lack of examples for practice. A set of " wrong " and " right " examples or a list of design rules cannot by themselves cultivate in the beginner the habit of methodical planned thinking.

The advantage of working to a properly directed plan lies mainly in the avoidance of all superfluous repetitions. The man who pursues an accidentally discovered solution without considering the consequences will often find that he has strayed into a blind alley and must start again at the beginning. Only by working to a methodical plan can the designer hope to escape unwelcome surprises of this kind. By adopting the right method of working and thinking carefully about it he can save time, avoid wasteful mental effort and thereby increase the effectiveness of his work.

One point must be emphasized without delay. Anyone who imagines that working to a method is a welcome opportunity whereby even a subject like design can be learned with minimum outlay of mental effort and without independent thinking will be quickly and profoundly disappointed. A methodical plan of working does not offer a substitute for intellectual abilities like imaginative power, logical thinking, concentration, the gift of combining ideas, and an inventive mind. It only points the way.

II. THE DESIGNER

Qualities required in a designer

Every student who wants to become a designer should bear in mind that design calls not only for absolutely clear-cut and purposeful intellectual activity, but also for an inventive and intuitive mind allied to a whole series of character-based and personal qualities. These qualities, however, are not capable of being acquired, but have their origin in a special endowment of the individual.

The following list gives a survey of the capabilities and qualities needed by the successful designer.

1. Capacity to visualize bodies, static forces and stresses, dynamic phenomena, hydraulic forces and flow conditions, electrical and thermal phenomena.
2. Integrating capacity.
3. Ability to think logically.
4. Ability to concentrate.
5. Inventive talent.
6. Memory.
7. Conscientiousness.
8. Sense of responsibility.
9. Integrity.
10. Perseverance.
11. Strength of will.
12. Aesthetic sense.
13. Temperament.
14. Personality.
15. Ability to speak and write skilfully.

1. *Capacity to visualize*

A well-developed capacity for visualizing is one of the basic requirements of the engineering profession, and particularly of the designer. His creations are always bodies composed of the simplest possible basic forms, such as right cylinders, cones, and spheres which he shapes, works upon, and assembles in his mind before putting them down on paper in the form of drawings. The designer must also have the imaginative resources to appreciate the interaction of components, the transmission of forces through them, the distribution of internal stresses, and all the physical phenomena occurring in a machine or piece of equipment.

Naturally, there are different degrees of this ability. Even of a beginner, however, it must be expected that he will at least have the ability to imagine simple basic forms and their combinations, interpenetrations and sections. Those who find it necessary even at this stage to use models to assist their imagination will never reach the status of the independent designer. Even the engineering draughtsman needs a certain imaginative power.

2. *Integrating capacity*

The capacity to visualize and the capacity to integrate are major constituents of a creative imagination for which the designer must have a certain natural aptitude. All machines and industrial products consist of known basic structural elements. By combining these elements the designer is continually creating new forms to serve specific ends, even when there are no pre-existing designs to guide him. It is also an established fact that by suitably combining existing inventions it is possible to evolve something entirely new which is in its turn patentable. Only by the skilful exploitation of natural laws can the designer make the effects of the laws serve his plans.

3. *Ability to think logically*

The intellect must be freed for concentrated productive thinking by eliminating to the fullest possible extent all unprofitable intermediate tasks of secondary importance and all distracting influences. This calls for the possession of highly developed intellectual powers on the part of the designer. He must be able to judge correctly the interrelationship between cause and effect, and to distinguish essentials from non-essentials. His judgment of the nature and magnitude of the various influences resulting from the different factors involved in a technical phenomenon must be straightforward and clear-cut.

A point which must be given special emphasis at this stage is that in only a part of his deliberations and decision-making can he call on the assistance of mathematics. His intellectual activity often consists in the abundant use of ordinary clear-sighted common sense.

The possession of this natural gift is therefore the main factor in deciding the extent of a designer's capability to reach the right solution to a variety of problems, to find means of making improvements, or to indicate new and improved ways to attain a specific goal.

4. *Ability to concentrate*

All successful intellectual activity calls for exclusive pre-occupation of the individual's entire thinking capacity with the problem on which he is engaged. Design thinking likewise demands very intense concentration at a high level. The necessary capability for this can only be acquired by long practice. Nervy, excitable, and restless individuals never learn the art.

5. *Inventive talent*

Most people regard an inventor with a certain awe. They imagine that the object invented is a kind of sudden revelation manifested by a special intuitive talent. Of course no one would deny that inventing calls for a certain natural endowment. But this consists in the inventor's ability, based on clear logical understanding, to advance stage by stage by judging, deducing, and combining until he achieves something new, an invention in fact, although in some circumstances he may not be able to recollect the process by which he reached his goal.

Reuleaux has noted that, " in inventing, one idea continually gives rise to another so that a veritable step-ladder of ideas is negotiated before the objective is reached.—There is no evidence of inspiration or flashes of illumination."

Inventing is thus a systematic intellectual activity, and it is therefore equally possible to speak of a methodology of inventing. It follows, too, that up to a certain level inventing is teachable.

Every designer, of course, needs some inventive capacity to call on when looking for possible solutions to a specific problem or combining familiar mental images to form a new product. If the inventive spirit is made to serve rigorous purposeful activity in the design field, it is to be welcomed without qualification.

There are, however, designers who appear to be obsessed with inventing and who are constantly putting forward new ideas. A special warning is needed against this sort of passion. It is only very rarely that the inventor derives any financial success from it. Krupp has said that " a good designer finds it easier to move, through the fruits of his labour, from the garret to the drawing room than does an inventor. The latter usually lands out of the drawing room into the garret."

6. *Memory*

Like all who work with their brains, the designer also needs a memory of average capacity. In the first place, of course, he needs this for studying the underlying sciences. In addition, part of the mental equipment of the designer consists of a vast amount of facts and figures which he must have at his finger tips all the time without needing to consult books. A good memory also helps him over a period of time to amass a store of experience which will be of value to him in later design work.

No less important than his intellectual capabilities are his personal characteristics. Despite their importance in his later professional life, it is unfortunately just these qualities which receive so little training and observation during the designer's student period. The beginner might therefore gain the impression that the qualities which make up his character are not so very important. However, there are many who have blundered through lacking these qualities and who have found the greatest difficulty in retrieving their lost confidence.

7. *Conscientiousness*

One of the principal characteristics, and one which can rightly be demanded from a junior draughtsman, is the ability to work thoroughly and conscientiously. The smallest error which finds its way from the drawing office into the shops can cause very serious harm under modern conditions of batch or mass production. It can also happen, however, that certain of the designer's oversights, such as unsuitable choice of material, or insufficiently generous dimensioning of parts, fail to make themselves apparent in the production shops. This sort of thing is even worse, because complaints from customers are harmful to a firm's reputation. A designer who makes mistakes of this kind soon forfeits the prestige he enjoys.

8. *Sense of responsibility*

An independent designer lacking the courage to accept responsibility is unthinkable. Courage of this kind springs from the self-confidence which a designer possesses when he has complete mastery of his subject. He who lacks the inner compulsion to acquire intellectual independence and assume responsibility had better give up any cherished ideas of professional advancement.

9. *Integrity*

Young designers are usually lost in admiration of the outcome of their first efforts at design and are therefore quite disheartened when corrections are made to their work. Integrity towards himself demands from the designer that he shall also have the courage to be self-critical of his work which, after all, is to be considered as no more than an approximation to the ideal solution and therefore always capable of still further improvement. When judging the work of others, however, it is best to refrain from criticism if one is not in a position to offer a better solution.

10. *Perseverance*

It must be admitted that even in the field of design there are many tasks which are not in themselves of absorbing interest as mental exercises, and which for this reason are considered boring. Instances of these are the calculation of the weight of the many components which make up a vehicle, or the determination of the position of centre of gravity. One should remind oneself, however, that even tasks like these must be performed for the sake of the design generally; this will induce the right attitude of mind and the perseverance needed to cope with them.

11. *Strength of will*

There is not a single designer who would not give thanks for his professional success to his exercise of will and to his powers of initiative and enterprise. Many examples in the history of engineering confirm that it was the strong-willed engineers in particular who achieved success and recognition in the face of all the objections and opposition they encountered.

12. *Aesthetic sense*

It has been said that everything which fits its purpose looks attractive. This, however, could mislead one into thinking that all one need do to obtain beautiful and attractive forms is to design to suit the purpose concerned. Although this is largely true, there remain unfortunately plenty of instances in which the designer must also rely on his aesthetic sense. On these occasions the designer with a marked sense of aesthetic values will benefit greatly in his work.

13. *Temperament*

As stated earlier, an overwrought nervy individual is no more suited than the phlegmatic type for an occupation like designing which calls into play qualities of intellect and character, as well as personal attributes. What is needed in a designer, therefore, is a harmonious and balanced temperament.

14. *Personality*

A designer occupying a position as section leader, departmental head, or chief designer, and therefore senior to many others, needs a quality which is taken for granted in every salesman, namely a positive presence and skill in dealing with the people he meets professionally owing to the important position he holds. He also needs some ability to judge character, so that he will be able to put the right man in the right job in his office and thus ensure fruitful cooperation.

15. *Ability to speak and write skilfully*

It is perhaps because of the quiet intellectual nature of their work that one so often meets designers who are unable to present their views fluently when the occasion arises. And it is the most capable ones who find, time after time, that their far-sighted and progressive work often runs into the most violent opposition. The designer who wants to make his views prevail in this situation must be able to apply to the task all his skill in speaking and writing.

What is a designer expected to know?

Every brain worker needs to have a certain store of knowledge for use in his job. Considered by itself this knowledge would have very little value. Only in conjunction with ability, systematic logical thinking, and the power to combine, judge, and deduce does it provide him with the means to do successful work.

For the designer, too, the information which he has accumulated in the various areas of knowledge forms the essential basis of his professional activity. What, then, are these areas of knowledge? For practical purposes the disciplines involved are the ones he acquires during his studies, ranging from mathematics to economics and management studies.

The following list gives a guide to the areas of knowledge of primary importance to the designer.

1. *Mathematics:*	Elementary and higher mathematics
	Descriptive geometry
	Mechanics: Solids (statics, strength of materials, and dynamics)
	Liquids (hydrostatics, hydraulics)
	Gases (aerostatics, aerodynamics, thermodynamics)
2. *Physics:*	Electricity
	Light
	Sound
3. *Chemistry:*	Inorganic and organic (fundamentals)
4. *Technology:*	Properties of materials (physical and chemical)
	Manufacturing processes (non-cutting, cutting, short-run and mass production)
5. *Theory of machines:*	Machine drawing
	Machine components
	Prime movers
	Mechanism
	Power transmission

The first stage in a designer's training consists in the acquisition of the knowledge and information whereby these disciplines are imparted. He will only derive value from them, however, if he continues to work upon the subject matter under the stimulus of questions and problems posed by himself until he has struggled through to sovereign mastery in the various fields. Arrived at the second stage of his intellectual development, he now also recognizes the great extent of relationships and inter-dependencies, and realizes that all the disciplines form an organic whole in so far as his profession is concerned.

It is now time to discuss some of the factors which are important for the young designer.

1. *Mathematics*

The big advantage gained by the user of mathematics is that the subject teaches the habit of systematic and logical thinking. For the designer, however, mathematics takes on a special importance, for it forms the foundation of many other special areas of engineering science. Mastery of mathematical laws and operations provides the mental equipment needed for investigating the laws governing the various physical quantities, and for applying the knowledge gained in this way to the solution of the designer's problems.

The next point, which has already been made on a previous page, is this. It must not be expected that all design problems can be solved with the aid of mathematical concepts and procedures. The beginner is very liable to fall into this error. One notices constantly that beginners starting work on a design problem search eagerly for formulae which will provide the solution, instead of first giving their common sense a chance to speak. Design problems which can be dealt with by the use of a certain mathematical formula, such as a design for a flywheel or for the blading of a fan, are therefore just what the beginner wants. The rising young designer soon discovers, however, that problems permitting satisfactory solution by calculation are comparatively rare, and that often it is just the problems presenting the greatest difficulty which are not amenable to mathematical treatment and have to be solved by mental activity of another kind.

For a detail designer or head of section a knowledge of ordinary higher mathematics is usually sufficient. On the other hand, a designer working independently and obliged to include the study of modern research work in the scope of his activities is forced to enlarge his mathematical knowledge accordingly.

Descriptive geometry.—The basic pre-condition for all design activity is, as mentioned previously, a good capacity for visualizing in three dimensions, and this remains true no matter whether the problem involves solid bodies, kinematic relationships, the action of forces, the distribution of stresses or fluid-flow phenomena. This capability, which, to a certain degree, must be inborn in the designer, can be developed by systematic work. A good way to start is by doing exercises in the perspective or axometric representation of bodies, the use of orthographic projection with front view, plan, and side view being introduced later or at the same time.

There are times, of course, when the designer resorts to a model for practical assistance. The usual reason for doing this is to clarify very complex three-dimensional layouts which raise problems of accessibility or feasibility of assembly in some already highly compact mechanical unit. Attempts to solve problems of this kind by graphical methods are often futile. A familiar example, of course, is provided by the automobile industry where models are used in order to give the fullest possible impression of the aesthetic aspect of the body design.

The " reading " of technical drawings showing complicated layouts is something to which considerable time must often be devoted before a clear idea of the object portrayed can be formed. This is why many firms seek to aid understanding by adding to the working drawing an axometric view to enable the men in the shops to form an immediate picture of the item concerned.

2. *Physics*

" The whole of engineering is only applied physics." These words indicate the importance of the subject in the professional activity of the engineer. The branches of physics of special interest to the engineer in general and to the designer in particular have developed into specialized forms for engineering purposes. These subjects are dealt with in special lectures which cater for the work which the designer will subsequently do. The subjects concerned are the mechanics of solid, liquid, and gaseous media, electricity, etc. An important point is that the engineer intending to take up designing should not only be familiar with the laws, but should also make appropriate allowance for them at the right stage in his design work. Experience shows that this is not an easy matter and that it calls for an intense appreciation of physical phenomena. Since one can usually observe only the effects and not the causes, it is necessary for the designer to form clear mental images of concepts like mass, force, inertia, friction, spin, thermal conduction, so that he can successfully tackle the task awaiting solution.

3. *Chemistry*

There are some engineers who attach little importance to chemical knowledge. They take the view that a designer only needs to know about the physical properties of construction materials and the various ways of working them. However, the engineer must also know about the structure of materials, their chemical behaviour, and their aggregate changes. And it is for this reason that he needs a knowledge of the fundamentals of inorganic and organic chemistry.

4. *Manufacturing techniques*

One of the most important elements in the designer's training is the study of manufacturing techniques. Alongside the important knowledge he needs of the chemical and physical properties of construction materials, the beginner must also familiarize himself with manufacturing methods and all the aids thereto. It is a recognized fact that the beginner should accumulate some experience in this field during his practical training before he starts his studies. Experience gained in this way, however, is not sufficient for design purposes, since during this early period the student will usually lack the necessary scientific basis for a deeper understanding of technological considerations and processes. In addition, the manufacturing techniques and high-performance special-purpose machine tools serving mass-production ends are constantly undergoing further development and advancement. It is therefore essential that the designer should keep up to datౖ by continuous study of the relevant literature and discussion with staff in the shops.

5. *Theory of machines*

The subjects with which the designer is concerned are as follows:

Machine drawing	Theory of form design*
Machine elements	Lightweight construction
Kinematics	Design of prime movers and of driven machines

Machine drawing.—Machine drawing is an aspect of the designer's craft. Its relationship to creative design activity is rather like that of typewriting to authorship. Assuming a certain amount of good intent and industry and some imaginative power, anyone is capable of advancing to the stage where, by employing well-known systematic rules, he can produce a satisfactory drawing suitable for workshop use, provided that he is given all the information necessary for the purpose. This book assumes such a capability. Instruction in machine drawing is given in a number of good textbooks.

Machine elements.—Every industrial product, no matter how large it may be, consists of a large or small number of individual components, known as *elements*, on the proper design and coordination of which the action of the whole depends. On closer study it is immediately obvious that a large number of such elements continually recur in the same role, although of course the shapes given to them and the materials and dimensions used are determined to a decisive extent by the special features of the application concerned. Most of these elements can therefore be brought to a common denominator, so that all that is left is a comparatively small number of basic forms.

Knowledge of these elements is of the utmost importance to subsequent design activity. There is an extensive literature available to the designer

---

* The term *form design* is used here as a translation of " Gestaltung ". This word, and also the word " Konstruktion ", can be translated *design* but in German the connotations are different. " Konstruktion " is used in a general sense referring to the whole planning operations of a machine. " Gestaltung " refers to the design of a single machine member. It is desirable to preserve these distinctions of meaning, and the term *form design* is accordingly used throughout this book.

on this subject. Unfortunately most of these works concentrate on mathematical treatment and ignore the many factors to be considered in designing.

Kinematics.—Where new designs are concerned it is of the utmost importance to know all the possible solutions capable of providing a specific effect, so that the best one can be selected from them. Kinematics, and synthesis in particular, shows the designer ways and means of finding such mechanisms. Consequently he must devote special attention to this study.

Theory of form design.—There was a time when it was thought that the engineering student could be introduced to the mysteries of designing by teaching him form design. There is plenty of published work on this subject. Form design, however, is only a part of the designer's activity and the " theory of form design " on its own is therefore not a suitable way of acquiring a comprehensive knowledge of design work. Designing covers all considerations and measures from the placing of the order right through to the graphical formulation of the solution in a manner fit for presentation to the shops.

Lightweight construction is concerned with designing with particular attention to the weight factor. Questions of this kind can only be handled by a man who is already familiar with the whole range of tasks implicit in design activity.

Design of prime movers and of driven machines is a subject, so one would imagine, which ought to offer the opportunity of learning design in its full range and scope. In actual fact, however, the situation is unfortunately one in which only design exercises are carried out on the basis of existing examples, so that the student designer has no need to rack his brains about the kinematic layout or materials or manufacturing and design problems. All that he needs to do is to take some of the leading dimensions and scale the existing design up or down. It is obvious that by this sort of copying no one can ever learn to appreciate design considerations or receive the training needed to produce an independent designer. Even now, to the best of the writer's knowledge, there is not a single technical college any-where in Great Britain or in Germany which teaches the science of design as a single subject according to a systematic plan. Is it to be wondered at that industry complains about shortcomings in the training of designers? Not for this reason alone has an attempt been made in the chapters which follow to present a methodical work-plan for the designer illustrated by simple exercises.

Before concluding these remarks on the essential intellectual equip-ment of the designer, reference must be made to one further important factor. There are some areas of knowledge which are already fairly complete in themselves, such as mathematics, mechanics, dynamics, hydraulics, etc., at least in so far as they enter into design. The chemical industry, on the other hand, is constantly supplying us with new materials, and new and better production methods and machine tools are always being developed. Engineering is engaged in rapid development scarcely to be matched by any other profession. This means that if a designer were to content himself with what he learned at college he would very soon be behind the times. To keep up to date with engineering advances he must give his attention to technical journals and make a study of patents which concern him. He must also make it his business to apply for copies of catalogues and leaflets for information purposes, and to collect diagrams and notes regarding observations and new knowledge which he has gained at lectures and exhibitions.

III. DESIGN FACTORS

A rational working plan for the drawing office

The planned work of the designer should cover the whole of his activity in the drawing office:

1. His purely intellectually creative work, in other words the activity generally termed *design* and comprising the system of working discussed in more detail in the pages which follow.

2. The organizational measures adopted in the drawing office, such as the sequencing and distribution of design tasks so as to facilitate properly planned preparation of the work for the shops, combined with speedy completion of the order. This planning of the work cannot be carried out at college, since the conditions for it are lacking. An attentive beginner, however, will soon learn it in practice.

Overall design is best left in one pair of hands, or at least supervised by a single design engineer. The tasks involved in this, starting from the problem to be solved, are the decision to use a given layout, the choosing of appropriate materials and suitable manufacturing methods, and the designing of individual parts at least to a sufficiently advanced stage to ensure that no further difficulties will be encountered during full elaboration or detail work.

When the work reaches this stage the design engineer in charge splits up the further detailing work into individual assemblies.

The principles governing the sequence in which the various tasks are carried out are determined by component delivery times. Experience shows that castings take a long time to deliver, particularly if they are bought out and if there is the added complication of obtaining quotations from a number of firms. A similar situation may arise with large forgings such as crankshafts. It is therefore necessary to take care to ensure that parts of this nature are tackled first and designed to completion. The production shops of the designer's own firm will also need early information on materials, tools, gauges, jigs and fixtures, etc. It may also be necessary to release suitable machine equipment and labour for production purposes. Consultation with the works on manufacturing facilities and work preparation is indispensable. In economically run works it has, after all, long been the practice to make use of planned work preparation in so far as the subsidiary work necessary for the manufacture of an industrial product falls within the province of the works.

By planning the execution of these preliminary tasks through the drawing offices, the rapid fulfilment of an order can be supported most effectively and the smooth and trouble-free development of all the subsidiary work involved guaranteed.

In the most general case, therefore, the designer's work-plan includes the following sections:

1. Exact determination of the customer's requirements or of the problem posed.
2. Checking the order as to feasibility of carrying it out with the firm's own resources.
3. Supply of all data, preliminary work, etc., needed for carrying the design through.
4. Ascertaining possible solutions and choosing the best solution.
5. Methodical working out of the overall design.
6. Splitting up of the overall design into groups.
7. Detail design, taking priorities into account.
8. Consultation with works on production planning.
9. Production design of assemblies and components.
10. Consultation with works, and possibly with customer also, regarding suggestions for modifications.
11. Production of all necessary works drawings.
12. Inspection of the finished item and/or carrying out of bench testing and evaluation of test results.
13. Obtaining of customer's verdict on behaviour in service and possible suggestions for improvements.

Since it is only possible to discuss below the purely design measures, it is only points 1, 4, 5, 9, and 11 which are of concern to the student designer.

What are the factors influencing design?

A detail designer will have little difficulty in allowing for design factors, because he will be given so many data and general principles that he will be able to solve his task without having any insight into the factors determining the system as a whole. Naturally in the course of time he will become acquainted with new aspects of engineering, so that he will gradually become familiar with the whole range of the designer's work. However, the path from being a detail designer working to instructions to being an independent designer is a very long one, and only young people who are aspiring and persevering succeed in traversing it.

It is therefore necessary that gifted students should, even during their college days, be made familiar with the solving of design problems which they will not be dealing with in practice until they become independent design engineers.

When working through a design problem the beginner soon realizes that, apart from taking into account all the customer's requirements, he must also bear in mind a whole series of factors concerned with manufacture. These will arise in such abundance that he will not know how or in what order he is to master them. At the same time not all the requirements can be taken into account in equal measure. Often they are quite contradictory. According to A. Erkens the art of the designer consists "in prolonged checking, pondering, and compromising on requirements which are often quite contradictory until there appears—as the end product of numerous associations of ideas, of a network of ideas—the design."

Since the factors involved in new tasks may arise continually in fresh forms, it is easily understandable why engineers like Reuleaux and Bach should speak of an infinity of requirements. If the common features of many such factors are grouped together under collective headings such as, say, manufacturing methods or properties of materials, then there result, as shown by C. Volk, about thirty such points which influence design.

This in itself is a very big step forward, for the beginner can now see whether or not he has overlooked one of the important requirements.

The influencing factors can be classified in various ways. To give the student a quick survey of them it is proposed to adopt the classification of Dr. Wögerbauer who divides them into:

   I. Tasks imposed directly by the customer's requirements.
   II. Tasks imposed by problems of manufacture in the shops.

I. *Tasks imposed by requirements*	II. *Tasks relating to manufacture*
Required action	Working principle
Mechanical loading	Mechanical loading
Climatic influences	Quantity required
Chemical influences	Design
Mechanical environmental conditions	Material
Size	Condition of material
Weight	Coating material
Fitness for shipment	Material availability
Handling	Manufacturing method
Maintenance	Assembly method
Overhaul	Labour required
Economy of energy consumption	Machine equipment
Service life	Limits and fits
Reliability	Quality of surface finish
Operating cost	Jigs, fixtures and tools
Appearance	Gauges and inspection facilities
Delivery date	Delivery date
Quantity required	Cost of manufacture
	National standards, works standard items
	Material standards
	Scrap utilization
	Patents
	Use of existing products

The significance of individual factors may change very greatly with different tasks. The listing above therefore does not imply any specific ranking.

How can one classify these factors clearly?

The individual factors do not exist independently of one another. They can therefore not be treated separately when solving a problem. Even a beginner engaged in solving the simplest of problems soon notices that the factors which he has to take into account are all interdependent and that, indeed, several of them need to be considered simultaneously.

A distinction can be made between direct and indirect relationships of these factors; for example, the quantity required affects manufacture directly and design indirectly.

If one investigates the mutual relationship of the various influencing factors one finds that working principle, material, manufacture, and design are involved in most relationships. These four points therefore assume a special significance. They form group foci which make for much greater clarity in the processing of design work.

These four group foci are listed below together with their influencing factors.

I. *Working Principle*
   Action
   Mechanical loading
   Climatic influences
   Chemical influences
   Mechanical environmental conditions
   Size
   Weight
   Fitness for shipment
   Handling
   Maintenance
   Overhaul
   Power consumption
   Service life
   Reliability
   Operating cost
   Appearance
   Delivery date
   Quantity required
   Freedom from noise

II. *Material*
   Form
   Size
   Weight
   Mechanical loading
   Chemical influences
   Climatic influences
   Quality of surface finish
   Manufacture
   Cost
   Service life
   Quantity required
   Delivery date
   Material availability
   Scrap utilization
   Freedom from noise
   Reliability
   Material standards

III. *Manufacture*
   Form
   Material
   Appearance
   Quality of surface finish
   Climatic influences
   Available machinery
   Delivery date
   Jigs, fixtures, tools
   Gauges
   Fits
   Patents
   Quantity required
   Cost
   Overhaul

IV. *Design*
   Working principle
   Mechanical loading
   Material
   Manufacture
   Mechanical environmental conditions
   Size
   Weight
   Standard items
   Existing products
   Appearance
   Handling
   Maintenance
   Overhaul
   Surface finish
   Fitness for shipment
   Power required

(H 642)

These four group foci likewise stand in a mutual relationship with each other. For example, design affects material and working principle. The material can only be selected after taking working principle, design, and manufacture into account.

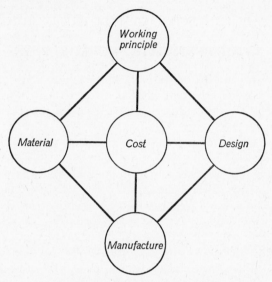

Fig. 3.—Showing how working principle, material, manufacture, design, and cost are interconnected

It will be seen, therefore, that design involves sorting out a very complicated set of relationships. Thus it is not surprising that the first effect of these observations on the beginner is very confusing and discouraging. Despite these complicated relationships between the design factors it is possible to organize the procedure and thereby to simplify it very greatly.

IV. A PLANNED POLICY FOR THE DESIGNER

A. THE SYSTEMATIC WORKING PLAN

Experience teaches that in all branches of human activity, no matter whether manual or intellectual work is involved, only a systematically planned method of working can guarantee success in the shortest possible time.

It is possible, of course, to advance arguments against the systematic method. There will be many, perhaps, who fear that a systematic working plan might force the designer into a Procrustean bed from which he can never escape. Anyone who believes this overlooks the fact that all those who do successful work, whether in the physical or intellectual fields, make use of such methods in their work—methods which they use consciously to begin with during the learning stage and then unconsciously as their mastery increases. Nor does the designer become hampered in his work, much less intellectually ossified, by using " systematic rules ". When methodically planned thinking has become habit with him he will be able to create all the more freely.

The systematic working plan is indeed time-saving, but not in the sense that one may regard design activity as piecework. The deliberations which necessarily accompany any successful design work can certainly not be accomplished in a fixed time in the way that is possible with a practised manual skill.

We have seen that the majority of the engineering factors can be accommodated in four groups. The only question now is to determine the most satisfactory sequence for dealing with them. Experience shows that the following path is the most practical one.

First, study the problem and prepare a freehand sketch or basic design to establish the required principle of action; then select the material, and finally carry out the design work with an eye to economy of manufacture.

That, in very general terms, is the systematic working plan for the designer. It can be illustrated diagrammatically on the lines of fig. 4.

Fig. 4.—The systematic working plan

26

We will now proceed, in the course of the pages which follow, to discuss these factors in the same sequence.

The overall working plan for the designer breaks down into the following sections:

I. Exact formulation of the problems and defining of all questions relating thereto.
II. Setting out of all possible solutions capable of providing the action called for in I in diagrammatic form (basic design) and selection of the optimum solution.
III. Selection of the most suitable material.
IV. Consideration of production engineering problems.
V. Deciding on the most appropriate form design.
VI. Ascertaining overall cost.

In its diagrammatic form the basic design represents a framework which gives only a hint of the final form. To arrive at a basic shape the designer must first make assumptions about the material and the method of manufacture, in other words he must run through the thought sequence

material → manufacture → form

After he has decided the basic shape, the designer has an approximate picture of the structure in front of him and can then proceed to a well-considered selection of the material. This done, and having taken into account all the economic factors for manufacture, he is also in a position to undertake the form design.

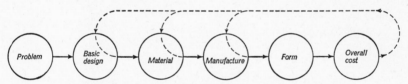

Fig. 5.—The systematic working plan when overall cost conditions
are not satisfied

Overall cost always decides the final form taken by the work. If the cost aspect is unsatisfactory it will be necessary to re-examine the situation (see fig. 5) in the sequence

material → manufacture → form → cost

B. THE PROBLEM TO BE SOLVED

Before a start can be made on the design solution of an engineering problem it is necessary to clarify as much as possible the various points which are to be taken into account when dealing with an order.

The types of problem to be solved

The demands which a customer may make can involve the following factors:

Action required
Mechanical loading
Climatic influences
Chemical influences
Mechanical environmental conditions
Size
Weight
Fitness for shipment
Handling
Maintenance

Overhaul
Economy of energy consumption
Service life
Reliability
Operating cost
Appearance
Delivery date
Quantity required
Freedom from noise

A universally valid sequence of customer's requirements in order of importance is not a practical possibility, since with each problem the importance of the requirements varies. In most instances, of course, only some of these requirements need to be fulfilled. For the designer, however, it is very important to have accurate knowledge about the priority ranking of the individual requirements. It will often be found that the customer himself is still unclear about them. It will therefore be of advantage for him to meet a representative of the firm so that he can define his requirements as closely as possible and be informed about the practicability of fulfilling them. Often, however, it is only the designer who can make a definite statement about the extent to which the customer's requirements can be met. It is always advantageous to both parties to define the requirements as accurately as possible prior to starting work on the solution.

For this purpose there are various systematic evaluation plans which can be used. Evaluation of the necessary requirements can be carried out on a points basis as follows.

1. Requirements which must be unconditionally fulfilled.......... 3 points
2. Requirements which must be fulfilled as completely as possible even though the solution leads to a compromise............... 2 points
3. Requirements whose fulfilment can only be insisted on if such fulfilment is economically practical........................... 1 point
4. Requirements which are unimportant and can be neglected...... 0 points

Here, as in many other questions arising in engineering, a graphical representation is preferable to a tabular one, and this is therefore given in the diagram (fig. 6).

Naturally, too much must not be expected from the graphical presentation. The scope of the problem to be solved will often be very large. The operating conditions and performance schedule for a locomotive occupy, for example, a whole book. For instances like this fig. 6 is unsuitable. But in many other instances, where the task is not so large, it brings the designer important advantages. He is able to spot immediately the difference in priorities to be given to the various requirements of the cutomer and, by drawing in the diagram lines for the solution reached, he will see right away to what extent he has satisfied the requirements.

The ideal would be to fulfil all requirements to the same extent, but usually it will be necessary to make do with a compromise. It must be insisted, however, that all requirements carrying two and three points are achieved.

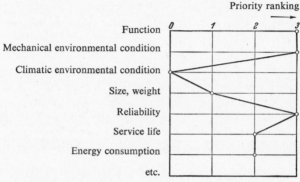

Fig. 6.—Characteristic graph indicating priority ranking for a given problem

We will examine below, in more detail, first of all the factors which in general meet the customer's requirements.

Function.—A specific function is required of every machine, every appliance, and every simple machine component. Depending on the action of the particular design element, so the following different types of function are distinguished:

mechanical, electrical, optical, thermal, magnetic, acoustic

Among these kinds of function it is really only the mechanical one which occurs on its own. All other components without mechanical function are also stressed mechanically in view of their mechanical construction.

It frequently happens that other functions come into play, as for example with commutator bars. These have the primary duty of acting electrically. Owing to the passage of current, however, they heat up and owing to centrifugal effects they are subjected to not inconsiderable mechanical loading. The fact should be carefully noted that these are actions which the customer does not want at all. The customer who orders an electric motor is only anxious to have electrical energy converted into mechanical power at a given torque and speed in the most economical manner possible. Apart from the specified functions, therefore, the designer also has to take into account a number of others arising out of the design.

The function called for by the customer is obtained by a meaningful arrangement of various constructional elements. It is usually possible to obtain the same function in a variety of ways. For example, if the problem involves making a work-table reciprocate, then there are several solutions which can be adopted. It now becomes the primary task of the designer to find out these solutions and then to select the best one in a manner which will be dealt with later.

Mechanical loading.—Apart from the loading imparted by the desired action, an engineering product is influenced mechanically by the place where it is used. These influences are forces, knocks, vibration, speeds and accelerations brought about by the particular conditions of use. Under this heading also belong all possibilities of damage arising from carelessness, such as jolts, allowing the product to drop, etc.

In most instances the customer will be unable to give any detailed information about mechanical loading of this kind. It is therefore the designer's duty clearly to identify and define it. If there is no empirical information available he must provide himself with the necessary data for calculation purposes by carrying out a theoretical examination or, if necessary, by experiments.

Climatic and chemical influences.—The immediate surroundings of the place of use have a powerful influence on design. The principal climatic effects are of a thermal and chemical nature. Depending on external conditions, therefore, the engineering product must be resistant to large temperature differences in summer and winter, to wide variations in humidity, to the chemical effects of a variety of gases and vapours, and to the corrosive action of water or aqueous solutions of salts, acids, bases, electrolytic cell formation, etc. These factors call for special attention in the design and construction of chemical plant.

The mechanical, climatic, and chemical influences of the environment naturally are closely related to the various modes of functioning as shown in fig. 7.

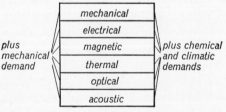

Fig. 7

Mechanical environmental conditions.—Under *environmental conditions* are grouped all the customer's specifications relating to the observance of specific location limitations. Such conditions can often cause the designer difficulty; for example, he may have to accommodate an overhead travelling crane in a very restricted space or find room for a gearbox in a very cramped location. In the automotive and aircraft fields considerations of this kind frequently confront the designer.

Size and weight.—For economic reasons alone, every designer endeavours to keep the dimensions of the product he designs down to the minimum, even though no restrictions in this direction are written into the contract. The demand that the amount of space occupied and the cost of materials used shall be kept to the minimum assumes such fundamental importance in design work that designers frequently see in these factors alone a guarantee of a correct solution to the problem in hand.

When, as in the automobile, aircraft, and shipbuilding industries, however, special rules are made regarding dimensions and weights, then the designer must make still greater efforts to satisfy the demand for minimum size and weight. All the design factors concerned with achieving an exceptionally lightweight design are grouped under the concept " lightweight construction ", to be dealt with later. At one time these principles were considered of such importance in the training of designers that a separate branch of study entitled " lightweight construction " was introduced in the engineering colleges.

Fitness for shipment.—The form design of engineering products must also take account of the various methods of transportation used. Apart from fitness for shipment, the question of transportation cost including customs duties also plays a part. Large and bulky items will be subdivided by the designer to permit shipment. Even where small products are concerned, however, it may be necessary to consider fitness for transportation; instruments, for example, are often very susceptible to jolts and vibration. They can be made transportable by arranging for the locking or dismantling of certain components.

Handling.—The main point to consider in connection with the handling of a product is what type of personnel—skilled or unskilled—will be operating it. Similarly, the mode of operation also decisively influences the form adopted for the controls and other devices used in handling the product. The controls on a locomotive will certainly be designed on more massive lines than those of a heavy goods vehicle, and the latter in turn will look different from those of a smart motor car. Many handles, levers, and handwheels are standardized. When free to decide the form of these elements, the designer will be guided by the considerations discussed above. The psychological reaction of the customer must also be taken into account.

Maintenance.—Maintenance implies attention to the tasks which have to be performed to keep plant or equipment in running order. It covers tasks such as cleaning, lubricating, retightening bolts, and changing worn or defective parts. To enable these operations to be carried out more quickly, the designer must arrange for easy access to the parts concerned and must limit the number of tools needed for maintenance purposes to the minimum.

Overhaul facilities.—As far as overhaul facilities are concerned, three possible situations must be considered. In the first place it may be found that to repair a simple product is more expensive than to replace the complete unit. In an instance like this it would be wrong practice to sub-divide further for the sake of replacement. When parts are subject to a high rate of wear it is often advisable to design them with a view to replacement and to supply the customer with replacement parts which he can fit himself. A third possibility is that specialized knowledge is needed to effect replacement or repair. In an instance of this kind it will even be necessary to take steps to ensure that the customer does not interfere with the product, but sends it in for repair to the firm that supplied it or to a suitable repair firm.

Energy consumption.—For financial reasons it will be of special interest to the customer that the product he is calling for should have the smallest possible consumption of energy, that is to say that it should be as economical as possible. As far as the designer is concerned this requirement means that in formulating the design of the product he must find out all the factors which have an important influence on the reduction of energy consumption, so that he can take them into account in his design. The economy of a product in regard to energy consumption is a reliable criterion of the quality that goes into its manufacture and is one of the factors determining product competitiveness with earlier and less advanced versions.

Service life.—Every purchaser of an engineering item is interested in securing the longest possible service life from it. It therefore remains an important principle for the designer, even if the customer makes no special stipulations in this direction, to ensure by appropriate dimensioning of parts, correct choice of material and attention to any other points, that the product will stay in service for the maximum length of time. No doubt the ideal situation would be for all the design elements to possess an equally long life span. It usually happens, however, that certain parts have a shorter life than the unit as a whole owing to mechanical, thermal or chemical effects. If a prolonged breakdown is to be avoided in an instance such as this it is essential that the wearing parts should be designed for rapid replacement.

Reliability.—Dependability is a quality held in high esteem by every purchaser of an engineering product which is required to operate continuously. Trouble-free performance at a specified rating or capacity depends primarily on the reliability and certainty of functioning of all the design elements involved.

From this there devolves on the designer the responsible task of constantly examining even the smallest details of his work as it progresses to discover what the individual components are subjected to in the way of harmful influences, possible overloading, wear effects or other troubles, and what repercussions these factors have on product reliability.

Operating cost.—Quite closely tied up with the question of energy consumption is the factor of operating cost. For the most part this finds its expression in the efficiency of the item concerned. The designer must therefore take into account all factors which have a bearing on efficiency. Unsatisfactory handling and operation in service can cause wear and leakage leading to loss of lubricating oil, cooling water, fuel, vapour, gas, etc., which will increase the cost of operation to a marked extent. In the customer's own interest, therefore, the supplying firm will provide him with:

Operating instructions.—The purpose of these is to give the customer information on operation and on recognizing faults promptly and rectifying them. At the same time operating instructions give the supplier a certain amount of protection against unfounded complaints and support him in disputes.

Appearance.—The saying " everything which fits its purpose looks right " applies in general to the designer. But there are many industrial products which the public wants to see designed with a clean and smooth external shape. Where machine tools are concerned, great importance is attached nowadays to this kind of appearance. Since external appearance has little influence on internal construction in this instance, the designer is able to meet the customer's taste for the contemporary without difficulty.

Freedom from noise.—A machine which works noisily will in the long run be unacceptable to the owner himself. Still less is it permissible to annoy the surrounding neighbourhood with such a machine, for such annoyance would very soon lead to legal action. For the designer there are various ways of eliminating or reducing undesirable acoustic effects. A noisy machine must be mounted in a manner providing acoustic insulation, and a change of site often helps to solve the problem. Treatment on a theoretical basis is usually inconclusive owing to complications introduced by surrounding structures; here only experiment can provide the answer.

Delivery date.—Every order made is subject to an agreed delivery date. Sometimes the date is laid down by the customer, but usually it is the result of joint negotiation. A delivery date must never be fixed without prior discussion with the drawing office and works, particularly where new designs are concerned. When large orders are involved, compliance with delivery dates is often enforced by agreed penalty clauses.

Quantity required.—Information on the quantity required will of course always be provided. It is of great importance in settling questions of material and production method, and in this way it significantly affects the design itself. For example, if the quantity required is small, it will be necessary to consider whether casting is still economical or whether it would be better to go over to welding or forging.

Overall cost is of the greatest interest to the customer. It also influences the designer's work to the extent that he must endeavour to choose his material and design his product on economic lines, so that a competitive overall price will result.

Types of problem to be solved.—Two classes of problems are to be distinguished, depending on whether they originate from the customer or from the designer's own works, as shown by the following list:

Customer's orders
- Ready-made product
- Ready-made product with modifications
- Complaints about machines previously supplied
- Repairs
- New design
- Improvement

Works' orders
- Further development
- New design
- Invention

Customer's order.—The best policy is for the customer first to inform himself adequately through advertisements and recommendations of the existence of a firm which appears likely to meet his requirements. By making a direct approach to the firm concerned or to one of its representatives, he should then obtain expert advice for the purpose of exactly formulating and defining his order. Descriptive publications, leaflets, and catalogues should not only serve advertising purposes but should also contain all data of importance to the customer, such as principal dimensions, weights, guaranteed values, graphs where applicable, and simple explanatory sectional drawings. A properly informed customer will naturally be able to express his requirements in such a way that there will be no need for a lot of discussion with the drawing office.

The majority of orders will concern items which are in regular production but require slight modification in layout, or the addition of certain attachments to meet the customer's special requirements. Modification work of this kind forms the principal activity of the design office in a lot of instances. The work involved is mostly adaptive design.

Apart from design work of this kind, the customer may present parts to be replaced, repairs, or suggestions for improvements. This applies particularly where insufficiently tried new designs are involved. Adaptive design of this kind, however, calls for great care in the interest of preserving the firm's good reputation. The primary task of the designer in this situation is to examine the true causes of the complaint very carefully, so that the defects which have become apparent can be eliminated in future by appropriate modification and improvement.

Work which calls for the maximum of design capability is the solution of new problems presented by the customer. Very often it is only by experiment and theoretical investigation that the basis can be created for resolving the design side of a problem for which there is no existing solution.

Works' order.—Apart from customers' orders, every firm always has designing to do on its own account. Part of this work is concerned with improvements suggested by experience gained in after-sales servicing, and part of it is concerned with development arising from study by the firm of its own products.

A special class of order reaching the drawing office is the kind which arises from efforts made by the firm on the basis of a careful study of the market to create by its own efforts new designs or improvements in design which, backed by patent protection, will constitute a marked advance in the field of engineering concerned. In work of this nature the designer has to be specially mindful of economic principles, so as to create from the start a product which not only represents a significant improvement in the range of possible solutions, but also offers the chance of gaining an effective lead on competitive products by virtue of the price at which it is offered to the market. In this kind of design work the policy should be to create a product which, by means of interchangeable and adjustable parts, by the fitting of extras and special features, and by the provision of extension facilities, is able to offer the maximum of adaptability.

Exercise problems

Study of the design problem is part of the training of the young designer, and the problem must therefore be defined with sufficient detail and precision to avoid any need for further queries. To the best of my knowledge the colleges have not so far adopted exercises of this kind. The problems set are mostly so incomplete that the student has a great deal of latitude for making his own assumptions, and, if he is of a critical turn of mind, he will continually be asking his teacher questions.

In big firms the detail designer receives the information he needs for his work from a higher level, and he is usually not put in the picture regarding the full extent of the problem. In smaller businesses, on the other hand, where the functions of detail designer, section leader, and engineer in charge are controlled by one individual, he will soon reach a position in which he can set out the essential factors in correspondence concerning inquiries, or define a problem in definite terms by direct contact with the customer. This ability calls for special practice, however, and for this reason some examples are given below.

Problem 1

Exercise problems as set in colleges are often worded on the following lines, particularly when dealing with prime movers or driven machines:

Design the impeller, the volute casing, and the bearing pedestal of a vertical-type shaft pump required to deliver Q litres per second against a delivery head of H metres, when running at n r.p.m.

Does such a problem correspond even approximately to the circumstances found in actual practice? Need we be surprised if the student believes that the context of a problem ought always to be as straightforward as this, free of all further limiting conditions and special requirements? In the form in which it is reproduced here, the problem quoted above is nothing more than an exercise in calculation. All the rest is copying from tried and proven models or patterns.

The problem as it concerns us is to indicate, on the basis of the list of customer's demands, what further particulars must be given to clarify the order as completely as possible.

Problem 2

A customer has sent in an inquiry with the following details:

In a bay of a chemical plant four autoclaves are already in position (layout plan, fig. 8). An overhead travelling crane with an electrically-driven monorail crab having a capacity of 500kg is available for lifting a basket into and out of the autoclaves. Since this crane cannot be used at the same time for opening and closing the cover, it

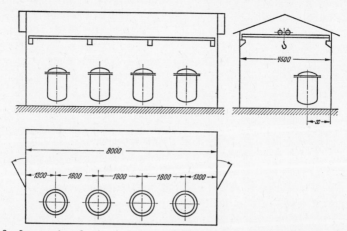

Fig. 8.—Layout plan of a chemical plant containing 4 autoclaves (dimensions in millimetres

is desired to install a device to do this job under the control of one man. The autoclaves are used for boiling and contain a 10% caustic soda solution. Boiling is carried
out at one-tenth atmosphere. The dimensions of the autoclaves are given in fig. 9.
The only items fitted in the covers are a pressure gauge and a thermometer. A sealing
washer stuck to the bottom flange prevents any escape of steam or vapour. In an
eight-hour period of duty the covers are opened at four-hourly intervals.

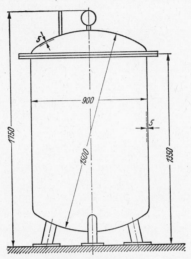

Fig. 9.—Autoclave

How long would it take to deliver the device and what would be the overall cost?
The designer should now make an examination to decide whether the information
given is sufficient to permit designing the cover lifting gear. This examination is best
carried out by referring to the factors which are basic to the customer's order. In
addition, the priority ranking of the individual requirements expressed by the customer
should be indicated in graphical form.

C. THE BASIC DESIGN

The designer's work starts with a study of the customer's requirements.
These were listed on page 28 and they were then discussed individually.
For the purpose of working out possible solutions, however, not all of these
factors are equal in importance. First and foremost, of course, is the
requirement for a specific action. The basic design is influenced also by
considerations of mechanical stressing, siting conditions, size, and economical energy consumption, whilst the other requirements affect choice of
material, manufacturing method, and form design, and must be considered
in these contexts.

Generally speaking, a number of solutions will be found for a given problem. The beginner would be well advised not to content himself with one or two such solutions, but to try to work out every possible answer in the form of simple diagrammatic sketches. The advantages in this for the designer are twofold.

1. He gains a feeling of confidence because he knows the various possible solutions as well as their advantages and disadvantages. Possessing this knowledge he can quickly counter any adverse criticism.

2. He protects himself against the unwelcome surprise of discovering that other firms have found a better solution to the same problem. On looking through patent specifications one notices over and over again how a patented design has been by-passed and even improved on by other firms. If the designer knows all the solutions to the problem concerned he can formulate the patent in a way which will eliminate all competition.

How are the various possible solutions found?

The discovery of a solution to the requirements raised by a given problem is only rarely the result of sudden inspiration—for all that inventiveness is demanded of the designer; rather is it the outcome of methodical study. In any case the designer should not rely on his intuition alone. The beginner must be disabused of the idea that this kind of work involves only inventive activity.

As long as eighty years ago Reuleaux also dealt with this question in his book on kinematics. He said that " in the field of machine problems (by which are meant mechanisms) the same intellectual operations can be introduced as are used also by science in conducting research in other areas ", and that " inventing is thinking, so that if we can organize the latter for our purpose we have also paved the way towards inventing."

There are various means by which the designer can work through to the discovery of solutions of a problem. It would therefore be a waste of time for him simply to wait until something suitable occurs to him.

The systematic designer will work methodically through the points listed below and in the process will be sure to arrive at several solutions.

The following sources will provide ideas leading to solutions which satisfy the problem.

1. Methods already established
2. Kinematics
3. Constructional elements
4. Historical development
5. Patent specifications
6. Technical journals and lectures
7. Exhibitions
8. Physical characteristics
9. Other specialist fields
10. Experiment

1. *Methods already established.*—Faced with a requirement for a certain mode of action, the designer's first and quite automatic response will be to search his memory for a solution to a similar problem. Often this method leads to success. The point which the beginner should note is the great benefit resulting from a good memory—a factor which has been mentioned earlier. Even during his student days he can consciously begin to accumulate the necessary experience by looking through catalogues and prospectuses, studying technical journals, visiting exhibitions, and observing life generally. While he is doing this it is necessary of course that he should reflect on the various problems involved and on the various mechanisms he meets, even if they are concerned with nothing more than the construction of a propelling pencil, and should attempt to re-create them.

2. *Kinematics.*—The study of kinematics is one of the prerequisites for the creative activity of the independent designer. Kinematics, however, is also a rich source of solutions to a very wide range of stipulated actions. The question of primary interest to the designer is of course: "What are the known possible solutions to a specific kinematic problem?" He will therefore start by looking for a collection of examples drawn from mechanisms already developed. Plenty of books providing this kind of information are available.

3. *Known constructional elements.*—Every piece of equipment, every machine, whether simple or complex, consists of the constructional elements dealt with in works on machine components. What could be more obvious, then, than to run through these elements in succession, and to consider whether and how they can be used to solve the problem. A simple example will illustrate this. The following are known:

Mechanical Elements

Levers, rods	Ramps
Screws	Ratchets
Wheels (toothed gears, friction wheels)	Couplings
Ropes, steel bands, belts, chains	Lifting and sliding valves and stop cocks
Springs	Cylinders, pistons
Cams	

Hydraulic elements
Pneumatic elements
Electrical elements
Optical elements
Acoustic elements

These constructional elements can of course be combined with each other to give a very large number of arrangements. When looking for solutions, however, it is advisable to start by considering the elements singly and asking, for example: "Can I build the mechanism with levers or with screw and nut?"

We will now consider a problem involving the conversion of a reciprocating straight-line motion into an equal motion displaced 90° from the first. At this stage no other requirements are to be met.

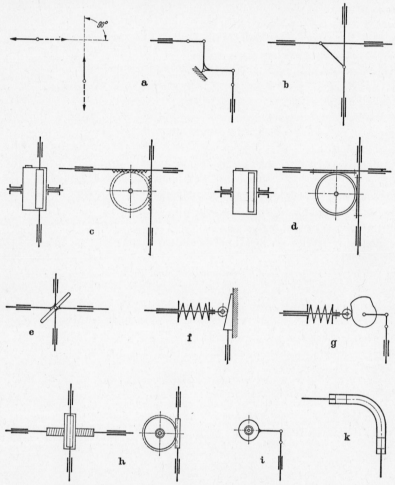

Fig. 10.—Survey of possible solutions. The function required is shown at the top left

(a) and (b) Solutions using rods and levers
(c) and (d) Solutions employing wheels
(e), (f) and (g) Solutions employing cams and ramps
(h) and (i) Solutions using screw and nut
(k) Solution employing hydraulic-transmission elements

The familiar solution employing a bell crank (fig. 10a) will immediately be recalled by every reader. One or other of the alternative solutions indicated may also be found from memory. We will try, however, to indicate further solutions by considering the available constructional elements. Solution *b* may be familiar from kinematic studies. A further solution (*c*) can be found by using a gearwheel and rack. For many applications a rack-type drive is replaced by steel bands wrapped round a pulley and attached to the latter and to the rod, as in solution *d*. Versions *e*, *f*, and *g* show solutions obtained with a slotted link, ramp, and cam.

The screw-and-nut principle combined with a rack or lever mechanism can also provide a solution as in versions *h* and *i*. A purely hydraulic form of transmission as in version *k* often presents many advantages.

It is not intended to suggest that versions *a* to *k* exhaust all the possible solutions. A combination of elements would certainly lead to still further solutions. The beginner will have noted, however, that the process of systematically working through the individual constructional elements constantly stimulates new " inventions ".

If no further conditions are laid down all the solutions considered meet the requirement originally stated.

A later chapter will show how to proceed in order to select the best solution.

4. *Historical development.*—A study of the historical development of a product provides very valuable pointers to the discovery of possible solutions. In technical literature, of course, there are few self-contained historical studies. The designer is therefore usually obliged to search out the data he wants from a variety of books and journals. This study of sources is very time-consuming and cannot always be expected of the designer. There is, however, a simpler way of studying the historical background, and this consists in looking through patent specifications in the subject concerned. Since these specifications are arranged in chronological order it is a simple matter for the designer to make a quick survey of existing designs. In this way it is possible for him to avoid the error of repeating existing solutions which provide the action required, because he will see to what degree of completeness problems of the same kind have already been solved and in what direction further development can be pursued.

5. *Patent specifications.*—The study of patent specifications thus provides protection against re-invention and gives a valuable stimulus to the discovery of new solutions.

6. *Technical journals and other publications.*—The experienced designer will never omit to keep up to date with the technical literature of his own field, and will keep written notes with an index and a brief indication of the contents. *Proceedings* of technical institutions, etc., can also give the designer many a valuable idea which will assist him in his work. The exchange of experiences with professional colleagues, even if active in other fields, is another method which provides useful ideas.

7. *Exhibitions.*—One of the main advantages of exhibitions from the designer's point of view is that they usually present the latest developments before they appear in print. The designer will also gain many a useful idea from being able to see and compare the products against which he is competing.

8. *Physical characteristics.*—A very worthwhile way of finding a new solution is by systematically going through the natural physical effects.

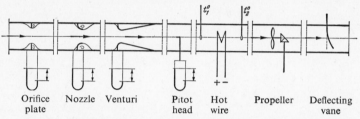

| Orifice | Nozzle | Venturi | Pitot | Hot | Propeller | Deflecting |
| plate | | | head | wire | | vane |

Fig. 11.—Methods of measuring gas velocity based on physical effects

If, for example, the problem is to devise an instrument for determining the velocity of a gas flowing in a tube, it will be found highly profitable to look for those physical quantities which vary directly or indirectly with gas velocity (fig. 11).

In this way all the possible methods of measuring gas velocity are found.

9. *Other specialist fields.*—As mentioned earlier, the training of a designer should be made as broad as possible, because by observing work in other specialist fields he will often pick up valuable ideas suggestive of new solutions. The motor-car designer, for example, may find that he is able to make good use of many an idea borrowed from locomotive building or aircraft manufacture.

10. *Experiment.*—New solutions to problems are often found by experiment. Usually this method is employed to realize some action or effect for which the necessary basis of theory or of practical experience is lacking. Random experimentation of course cannot be tolerated in work of this kind. Instead, the experiments must be carried out in accordance with a fixed and pre-arranged plan if they are to lead to success.

How is the best solution found?

The problem given on p. 40 originated as a detail in an exercise devoted to the design of a steam engine. The problem was presented in the following way:

Find the best mechanism to operate the inlet valve indicated in fig. 12 which moves horizontally in the cylinder head of a vertical high-speed two-cylinder uniflow engine working on superheated steam, delivering 20h.p. and running at 800r.p.m. Eccentric throw and valve travel are 30mm.

The generally known solution to this valve gear problem was the one employing a bell crank. One student, however, asked whether a rack-operated gear according to version *c* could not also be used. Stimulated by this question, some of the other students put forward further suggestions, so that in the end we systematically drew up all the possible solutions as shown in fig. 10 *a–k*. This set off endless criticism of the advantages and disadvantages of the different versions suggested. If there are just one or two requirements to be met, the problem of deciding which is the best

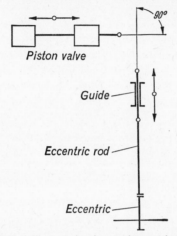

Piston valve

Guide

Eccentric rod

Eccentric

Fig. 12.—Diagram showing a valve gear layout

solution is still a simple one. In fact, however, more and more requirements were brought up for discussion so that eventually the picture became fairly confused. This kind of situation is typical of many similar ones which arise in actual practice. The difficulty here is not just to find a way of producing the specified action, but also to solve problems which first make their appearance during the designing of the machine. Faced with a situation of this kind, the designer himself must first draw up a list of all the factors which bear on the problem.

In the example under consideration, therefore, the various influencing factors were first ascertained and listed as follows:

1. Accuracy of action required
2. Positive drive
3. Simplicity of layout
4. Low rate of wear at joints and guides
5. Low energy consumption
6. Small inertia effects
7. Low manufacturing cost

The evaluation scheme

To enable a choice to be made when many requirements are laid down, there has been developed for practical use a method which quickly yields the suitable solution. All the solutions found and all the conditions which the problem requires to be fulfilled are set out in an evaluation scheme, and the various solutions are then awarded points according to their suitability.

Evaluation on a points basis can be made on the following lines:

> Very suitable............. 3 points
> Quite suitable............ 2 points
> Suitable as a last resort.... 1 point
> Unsuitable................ 0 points

A finer gradation using more points is quite practicable and provides a more accurate picture. The occurrence of a zero in any column means that the version concerned is unsuitable. The best solution is therefore the one having the largest number of points. If the best solution found is compared with the perfect solution, we obtain a picture which shows how closely the ideal has been approached. This gives rise to the idea of a merit rating expressed in the form of a ratio:

$$\frac{\text{Number of points awarded to actual solution}}{\text{Number of points awarded to ideal solution}}$$

Designs which have actually been built have been assessed for merit rating on the basis of the evaluation plan, in order to ascertain where the weak spot lies in the design and where improvement and further development might be tried.

The evaluation plan is certainly a very useful means of making a selection from a variety of solutions. It does not, however, afford a mathematically accurate picture of the true circumstances. Nor is this necessary. After all it is sufficient, and indeed marks a big gain, to be able by this type of evaluation to identify a small group of solutions which attain the largest number of points. If, for example, the two best solutions rate 20 and 21 points, it does not necessarily follow that the latter is the one to be preferred. We shall see later how to pick out the best solution from two or three which appear to be of equal merit.

To anticipate any objections, let it be said right away that naturally not every problem entails the working out of an evaluation plan. This would take up far too much time. The experienced designer will usually be able to manage without it. But for the beginner it is certainly a useful aid to selection and to the training of a critical judgment.

It might even be objected that different designers would differ in their evaluation of one or other of the factors involved, and would therefore disagree in the final result. Experience in this field shows, however, that despite the variety of views likely to be expressed in regard to the evaluation itself, the final solution turns out to be practically always the same when the experiment is conducted with beginners.

In the example we are considering, the experienced designer would find the optimum solution without the aid of the evaluation plan. We will draw up the plan, however, to give the beginner a simple example of the way it is used.

The choice made will naturally be governed by

1. Requirements expressed by the customer, and
2. Factors which are often either not raised at all by the customer or are only raised indirectly, but which nevertheless decisively affect the design.

These factors are as follows:

Mechanical factors	*Non-mechanical factors*
Loading forces	Thermal
Stress pattern	Electric
Shock	Magnetic
Vibration	Acoustic
Mechanical strength	
Inertia effects	
Stiffness	

The evaluation plan for the kinematic problem mentioned earlier looks like this (fig. 13):

Solutions	*a*	*b*	*c*	*d*	*e*	*f*	*g*	*h*	*i*	*k*	ideal
Action required	3	0	3	3	0	1	1	0	0	3	3
Positive drive	3	–	3	3	–	0	0	–	–	3	3
Simplicity of layout	3	–	2	2	–	–	–	–	–	2	3
Low rate of wear	2	–	2	2	–	–	–	–	–	2	3
Small inertia effects	3	–	2	2	–	–	–	–	–	0	3
Low energy consumption	3	–	2	2	–	–	–	–	–	–	3
Manufacturing cost	3	–	1	2	–	–	–	–	–	–	3
	20	–	15	16	–	–	–	–	–	–	21

Fig. 13.—Evaluation plan

This result is thus seen to agree with the layout used in practice and could also have been arrived at, as mentioned earlier, without using the evaluation plan. The gain which the beginner derives from using it is that he is made to think about other possible layouts and now realizes why no attempt has been made to adopt any other type of drive. This is where he contrasts with the designer who cannot trust his own judgment and who looks around anxiously for existing models which he can copy without further thought.

The designer is constantly faced with problems of this kind. For example, is it immaterial or just a matter of personal choice whether a plain bearing is used or a ball or roller bearing? Even when the decision to use roller bearings has been taken after reasoned consideration, there remains the question of selecting the right type of bearing and bearing arrangement. One need only consider the problem of mounting a bevel

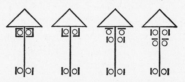

Fig. 14.—Bearing arrangements for a bevel gear shaft

gear shaft (fig. 14). How many possible arrangements can the designer choose from, and which one is most suitable? Or can it perhaps be argued that all the layouts shown are equally effective?

For the pivotal mounting of a hinged arm the designer has no less than eight possibilities to choose from. Which one should he decide to use?

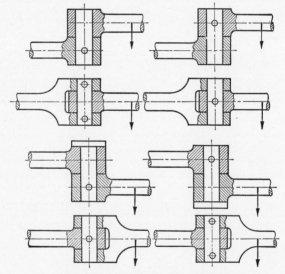

Fig. 15.—Variants of a hinged joint

(See fig. 15.) Those who have not learned to solve problems of this kind will fail miserably in the attempt.

Many unselfconfident designers are reluctant to undertake any re-thinking about mechanisms which have become sanctioned as it were through their absolute domination of the field. One need only think about the parallel-crank drive used on locomotives. To reject every other solution out of hand, however, would be sheer prejudice. Why, for example, are the transmissions shown here unsuitable for coupled wheels? (See fig. 16.)

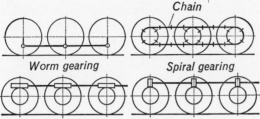

Chain

Worm gearing   *Spiral gearing*

Fig. 16.—Methods of driving coupled wheels

It is not necessarily always kinematic problems, however, which need to be examined to determine the advantages and disadvantages of the various solutions offered. An example of this is provided by a question

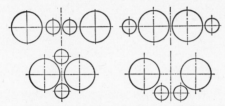

Fig. 17.—Alternative arrangements of twin cylinders and their associated valves

of the following kind: What alternative arrangements of cylinders and valves are possible in a vertical duplex steam engine? (See fig. 17.)

Similar examples will confront the designer all the time. But we must

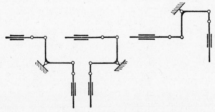

Fig. 18.—Various arrangements of a bell crank

now return to our original problem. With the bell crank arrangement (fig. 10a) adopted as the most promising solution, the next step is to decide on the basic form. Here again fresh questions immediately arise. Must the crank be positioned exactly as shown originally, or might it not be better to mount it in some other way? (See fig. 18.)

The general layout will settle this problem without difficulty. As is well known, valve gears are required to maintain a high degree of dimensional accuracy. If slackness develops in every joint, this will adversely affect the valve timing and, consequently, the economic operation of the machine. The idea might now suggest itself that some of these joints could be eliminated by using one of the arrangements shown in fig. 19. If one considers, however, that experience shows that pivots tend to wear less than slidebars and guideways, it will be seen that arrangements *a* and *b* must be avoided. Apart from this, these two arrangements are expensive and also absorb more energy. The problem originally set is now solved, and after the material and manufacturing method have been decided on, the work of designing the actual form of the mechanism can proceed.

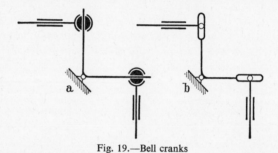

Fig. 19.—Bell cranks

There are some beginners in design work, however, who are so infatuated with their own solutions that they attach no importance to the result of the evaluation plan. For those who suffer from this trouble there is a simple remedy which can easily be applied in teaching, but is not always applicable in actual practice owing to the loss of time which results. The remedy is to let the sceptic develop the design by which he swears, which may be version *f*, *g*, or *h* (fig. 10), through all its stages. He will then very soon discover that the difficulties become greater and greater. A method often used in the past was to allow separate groups to work through the solutions found, and then to adopt the solution offering the biggest advantages. This, however, is very time-consuming and uneconomical.

What is to be done, now, if the evaluation plan produces two solutions with the same number of points? Are the two solutions really equal in merit? Only rarely is it safe to assume this. Even when this situation arises, it is still possible to reach a definite conclusion by taking a closer look at the solutions which have the appearance of equivalence. To obtain a final verdict it is usually sufficient to make a more precise formulation of the requirements to be met and to include some extra factors relating to production questions. It will often be found that the question of manufacturing cost alone decides the issue.

Exercise problems

The technique of designing can only be learned through constant practice. A methodical plan of working, design principles, criteria, and examples of " bad " and " good " practice will not by themselves lead to success. The student designer should not spend all his time just accumulating fresh knowledge and information; instead, he should in due course acquire an ability for creative working. This stage is only attained through a great deal of practice, and that is why this chapter ends with some typical exercises. The solutions to these problems will be found in an appendix at the end of the book (p. 235).

Problem 3

The straight-line reciprocating motion of a rod is required to be transmitted to a second rod parallel with the first in such a manner that the rods travel equal distances but in opposite directions (fig. 20). Find the possible kinematic solutions.

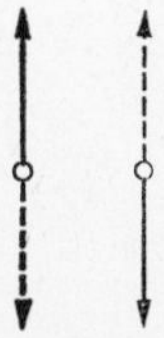

Fig. 20
Rod motions
required

Problem 4

A cast-iron table (fig. 21) 1200mm in diameter and 8mm thick is required to be given a parallel rise-and-fall motion without shifting sideways or rotating.

The conditions applying to this problem are as follows:

1. Assume that the effective load is 100kg and that it is applied as near as possible to the centre of the table.
2. The table is required to move vertically through a total distance of 100mm.
3. The table must have a perfectly smooth surface and be capable of being locked in any position.
4. The table base (cast iron) should be in the form of a circular column having an inside diameter of 300mm.
5. The operating mechanism is to be accommodated inside the table base.
6. The operating device shall be a handwheel with crank handle and horizontal shaft.
7. Fine adjustment to an accuracy of 1/10mm is required.
8. Ten of these units are required.

The first step is to investigate the possible solutions to the problem of providing the required motion, whereupon the evaluation plan can be used to find the optimum solution.

Problem 5

The pendulum saws at one time widely used for cutting planks to length (fig. 22) have since dropped out of favour owing to their many disadvantages, such as physical effort required, liability to cause accidents, confinement to a fixed location, or large amount of headroom required. Saws of this kind have been replaced by cross-cutting saws of low overall height which are free of the disadvantages mentioned.

The problem is to discover which mechanisms can be used to impart the necessary horizontal straight-line reciprocating motion to the circular saw.

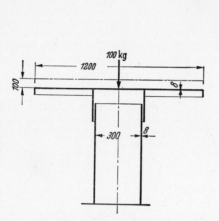

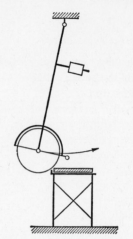

Fig. 21.—Dimension diagram
for a work table

Fig. 22.—Schematic diagram
of a pendulum saw

Problem 6

In the course of designing a worm gear unit having the worm on top, a designer runs into the problem of where to split the housing.

In how many different ways can this be done, and which way is the most advantageous?

Problem 7

The idea occurred to a student designer working on the design of a screw jack that, in addition to version *a* (fig. 23), another version illustrated at *b* would be a possibility. Why is version *a* usually preferred? Give a short account of the advantages and disadvantages of methods *a* and *b*.

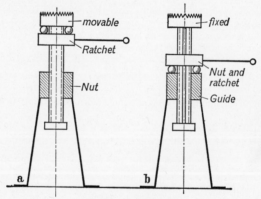

Fig. 23.—Screw jacks

D. MATERIALS

Which factors determine the choice of material?

The choice of the most suitable material for a component is one of the most difficult the designer has to face, one of the complications being the very large number of factors which have a bearing on the problem. In formal instruction on design it is just this problem which is treated lightly. Often nothing more is done than to select the material from a table giving the type of application for which it is suitable. Many a time the beginner is given a free hand in this direction because no specific requirements have been laid down. In the practical field it is not so many decades ago that the problem of selecting a suitable material could be accomplished without much difficulty. The times are past, however, when bills of material merely specified cast iron or steel. Modern technology has created an abundance of materials possessing the most diverse properties. In steel alone, there are over a hundred unalloyed grades and several hundred alloy grades, and the development of new materials continues all the time. What new prospects have been opened up to the designer through high-alloy steels and aluminium alloys. As far as the application of moulded plastics, for example, is concerned, we are still at the beginning. What a range of applications presents itself even now, thanks to the heat resistance and impact strength of synthetic resins. A stage has already been reached where plastic materials based on synthetic resins are competitive with steel fabrications (viz. automotive body building). It is safe to assert that in many fields the advancement of design has become purely and simply a question of finding suitable materials. The further development of the steam turbine, for example, would have been out of the question in the absence of materials offering high strength at elevated temperatures. The same applies to an even greater degree in gas turbine construction.

The factors determining choice of material have already been listed on page 24, but are repeated below:

Properties	Quantity required
Form	Delivery date
Size	Material availability
Manufacture	Scrap utilization
Cost	

The general heading " Properties " above covers a whole series of further factors, as follows:

Weight	Service life
Mechanical loading	Freedom from noise
Climatic environment	Reliability
Chemical environment	Material standards
Quality of surface finish	

It is time now to turn to a brief discussion of the principal factors affecting the choice of material to be used.

Material properties.—Only by using some material can the designer give tangible form to the mechanism he is called upon to produce. The material selected must satisfy all the requirements stated above; in other words it must possess the right properties for the proposed application. Indeed its properties decisively affect its selection.

Table 1 lists the principal properties concerned.

TABLE 1.—Properties of Materials

Physical	Mechanical	Technological	Chemical

Specific heat
Melting point
Thermal conductivity
Coefficient of thermal
  expansion
Electrical conductivity
Intensity of magnet-
  ization
Saturation
Remanence
Coercive force
Permeability
	Specific gravity
	Tensile strength
	Compressive strength
	Shear strength
	Bending strength
	Fatigue properties
	Torsional strength
	Buckling strength
	Elastic limit
	Modulus of elasticity
	Impact test figures
	Coefficient of
	  expansion
	Hardness
	High temperature
	  properties
	Wear resistance
	Sliding properties
		Castability
		Forgeability
		Capacity for
		  being rolled
		Deep drawing
		  capacity
		Machinability
		Weldability
		Suitability for
		  brazing
		Shrinkage
		Surface pro-
		  perties
			Resistance to:
			  Acids
			  Bases
			  Oxidation
			  Water
			  Oil
			  Greases
			  Petrol
			  Soap solutions
			Electrolytic cell
			  formation

In what follows the main factors involved will be briefly set out.

The type of material required can sometimes be quickly decided. For instance, a certain application may call for cooling or heating tubes which are required to have very good thermal conductivity. For duty of this kind there are available, of course, the ordinary commercial types of brass, copper, and aluminium tube. Consideration of the space factor or of the economic aspect of the problem will then quickly lead to a decision.

The principal materials used in mechanical engineering are cast iron, cast steel, steel, and light metals for parts subjected to fairly heavy duty. When very large forces are involved the material usually indicated is steel. Of course there is still the problem of deciding which grade to use, and this will be discussed in the next section. Choosing the material becomes much more difficult when the item concerned is a small piece of apparatus or equipment the strength factor of which is largely guaranteed by the dimensions conferred on it by its technological function. If, for example, the problem concerns the frame for a small appliance, then the materials which can be considered are cast iron, steel, bronze, red brass, brass, an aluminium or magnesium alloy, a zinc alloy, moulded plastics, or even ceramics. Any final decision as to choice of material must therefore depend on the other factors considered.

When the customer specifies a definite weight for a certain product, this is usually because he wants lighter weight than is given by a standard type of construction. It would be wrong to imagine that this objective can only be achieved by using a material of lower specific gravity. The methods available to the designer to enable him to achieve lightness of weight in his designs form a subject termed " lightweight construction " and are dealt with in detail under that heading.

Other criteria affecting the choice of material are coefficients of friction, wear and corrosion resistance, and thermal characteristics. Cast iron, for example, is characterized by the fact that " growth " sets in at temperatures above 300°C. Machine casings exposed to temperatures exceeding 300°C are therefore made of cast steel which does not exhibit this particular behaviour.

When a designer is asked to design an engineering product for a long service life, he usually tries to meet this demand by dimensioning to give reduced loading. It is quite likely, however, that other characteristics of the material will have a big effect on the life of individual components, possibly its resistance to thermal and chemical effects, and to wear. In instances where service life is subject to special agreement, the designer should select the materials used in such a manner that all parts are able to remain in service for an equal length of time, as far as possible, without requiring to be replaced.

So long as only the basic design exists as a schematic drawing it will rarely be possible to predict what type of material is likely to be used. For lattice-type structures, and for rods and levers in the larger sizes, there will be no question of using cast iron, for example. Under conditions of this kind a more definite choice of material is not practicable.

When welding was still in the early stages of its development one often heard the view expressed that only castable material could be used for complicated parts. Today, however, this no longer applies, since even very complex components can be fabricated by welding. Naturally, there are always other factors which affect the choice of material. Only by considering all the demands, including those involving other factors, can the final decision be taken.

Size.—Large items such as machine frames are nowadays usually fabricated in steel by welding. On the other hand, cast iron also is used for machine frames ranging from the largest and heaviest types (e.g. top and bottom sections of turbine casings weighing some 50 tons) down to castings used in the manufacture of small machines and weighing only a few pounds. It can be seen, therefore, that size alone does not decisively affect the choice of material used.

Manufacture.—The manufacturing aspect affects the choice of material in so far as special circumstances, or the demand for the lowest possible manufacturing cost, may indicate a specific manufacturing method. For example, surface qualities which are only obtainable through additional finishing operations or by the application of suitable coating substances may demand the use of special materials.

Cost.—In nearly all design work the question of cost plays a vital role. The main item is material cost and this depends very largely on material availability. It is sufficient to give the student the minimum of necessary data (possibly in the form of comparative prices) so that he can form a purely qualitative view of the cost position when choosing materials. The designer engaged on a design task in actual practice naturally cannot make do with this method. Instead, he must make his decisions on the basis of accurate calculation.

Quantity required.—The customer always states a certain figure giving the quantity of items he requires. This information provides an important lead in eliminating some of the possible materials. For example if only five units of a certain component are called for, it will scarcely be a practical possibility to cast them if the design is a new one, and instead the designer will consider using steel which can be fabricated much more economically by welding or hand forging.

Delivery date.—Here too, similar considerations influence the choice of material. A very early delivery date will often preclude manufacture in castable materials, and instead the designer will consider using steel.

Material availability.—For economic or political reasons it may prove to be very difficult or even impossible to procure certain materials. In a situation like this the designer is obliged to make do with other materials which may not always be a perfect substitute for the material which has ceased to be obtainable. Another difficulty in such times is the long delay and uncertainty affecting incoming supplies of materials, and this in turn often makes it hard to keep to the appointed delivery date.

How should one choose the material?

Generally speaking the customer does not express any wishes regarding the material to be used. When the order is for a public authority or another firm, however, this may not apply. In orders of this latter kind there may be extensive specifications to comply with in this respect.

It will be assumed now that no requirements regarding materials are stated in the problem as presented to the designer. The designer himself is therefore responsible for listing all those factors deriving from the customer's requirements which determine the choice of material. For instance, a unit required to stand in the open where it is exposed to all weathers will need to be resistant to atmospheric corrosion. If there is a possibility of mechanical damage through impact or shock, then clearly the material chosen must satisfy certain conditions in regard to impact strength, notch toughness, and elongation. Many of the mechanical requirements only manifest themselves at the design stage, and this applies to the various strength values ascertained by analysis of the forces involved. Before listing the different factors affecting choice of material, the beginner would be well advised to study Table 1 to make sure that he has not omitted one of the important points.

It is often possible for a component to be made in a very wide variety of materials.

A simple bracket, for example (fig. 24), can be made in cast iron, cast steel, malleable iron, steel either forged or welded, red brass, or aluminium alloy. Admittedly some of these materials will be dismissed as unsuitable

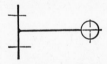

Fig. 24.—Bracket

as soon as detailed conditions are laid down. It will hardly ever be possible, however, to say at first glance which material is the most suitable one. The material most suited to the purpose will therefore be chosen, as a first approximation, by using the evaluation plan already discussed. The following example shows a typical evaluation plan as used for this purpose (fig. 25).

Material	Cast iron	Cast steel	Malle-able iron	Steel	Red brass	Alum-inium	Plas-tics	etc.
Mechanical strength Resistance to climatic effects Notch toughness Elongation etc.								

Fig. 25.—Evaluation plan

Continuing with the example of the bracket just considered, let us assume that the design problem runs like this:

A bracket is required which can be fitted to a tank standing in the open. In more detail, the following requirements have to be met:

Functions: The device in question is needed to support a shaft of 15 mm diameter.
Mechanical environmental conditions: 150 mm projection, mounting surface flat and at any height.
Mechanical stressing: Small forces involved, but possibility of damage.
Chemical environmental conditions: Exposed, outdoor location.
Service life: 10 years.
Maintenance: None.
Appearance: Smooth rounded shape.
Weight: Unimportant.
Quantity required: 20.
Delivery: 3 weeks.
Cost of manufacture: As low as possible.

The basic design in this instance is perfectly clear (fig. 24).

The short list of possible materials includes only cast iron, malleable iron, and steel, since cast steel is too good for the purpose, and red brass and the aluminium alloys are too expensive. Moulded plastic materials are unsuitable owing to exposure to climatic effects. We therefore arrive at the evaluation given below (fig. 26).

For the evaluation of merit on a points basis we will adopt for the present purpose a scale giving a finer gradation.

Very suitable..........	7 to 9 points
Quite suitable..........	4 to 6 points
Limited suitability......	1 to 3 points
Unsuitable.............	0 points

Material	Cast iron	Malleable iron	Steel	Ideal
Mechanical strength	8	8	9	9
Resistance to climatic effects	6	5	4	9
Service life	9	9	9	9
Appearance	9	9	7	9
Manufacturing cost	3	2	6	9
Number of points	35	33	35	45

Fig. 26.—Evaluation plan

This plan shows the cast-iron and welded-steel versions to be equal in merit. Since the factor of mechanical stressing is negligible, we will assume that ordinary engineering steel is chosen.

The lower manufacturing cost of the steel version has already been brought out in the evaluation plan. The detailed estimate shown in fig. 27 confirms this assumption. The manufacturing cost for twenty units is seen to be £12. 18s. for the cast pattern and £6. 8s. for the steel pattern.

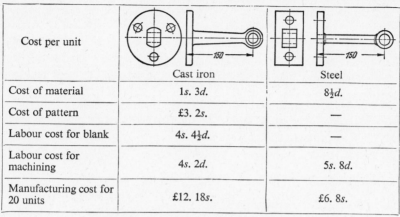

Cost per unit	Cast iron	Steel
Cost of material	1s. 3d.	8½d.
Cost of pattern	£3. 2s.	—
Labour cost for blank	4s. 4½d.	—
Labour cost for machining	4s. 2d.	5s. 8d.
Manufacturing cost for 20 units	£12. 18s.	£6. 8s.

Fig. 27.—Comparison of manufacturing cost

This is a quite unambiguous result and will lead to a decision to design the part as a welded steel unit, unless special importance is attached to the more pleasing appearance of the cast-iron pattern.

Selection of the right material, however, is not always as simple as in the example just considered. If the primary selection given by the evaluation plan suggests steel as the most advantageous material, the designer has still to decide which is the most suitable grade of steel. In this work he will find valuable information in data sheets which list the applications for which the various grades are suitable. It is necessary, however, to bear in mind the fact that each specialized branch of engineering, for example, hoist and crane construction, steel construction, automotive engineering, makes use of materials which it has specially selected for its own purposes on the basis of long experience. The designer must know all about these materials and turn to them first when making his choice. Only if they fail to meet his requirements should he decide to use some other material.

Principal materials used

The majority of the materials of principal interest to the designer engaged in general mechanical engineering are covered by standard specifications, but in some cases information can only be obtained from manufacturers' data sheets. Many handbooks include information about material properties. A list of British Standard Specifications dealing with material properties is given in Appendix C. (p. 255)

Cast iron

Cast iron is available in grades offering different qualities. Ordinary grey cast iron is adequate for general engineering castings for which no

3

special quality is laid down, that is to say for machine components of secondary importance. For more heavily loaded parts including, for example, highly stressed machine frames or cylinders, the high-grade variety is used. For exceptional conditions the designer can use special qualities which are acid-resistant, alkali-resistant, and heat-resistant, or the grade known as chilled cast iron. The strength properties of cast iron are conditioned by the rate of cooling and are therefore shown graded according to thickness in the standard specifications. The properties of cast iron are discussed in more detail on page 76 in connection with design problems.

Cast steel

Cast steel has greater mechanical strength than ordinary cast iron or malleable cast iron. It is therefore used when larger forces are involved and when casting offers advantages over welding or forging. Its properties are comparable with those of wrought steel.

Cast steel, of course, can also be alloyed. For special applications calling for extreme toughness, extreme hardness, extreme wear-resistance, etc., there is available a whole range of special grades, so that it is unlikely that the designer will have any difficulty when looking for a suitable cast steel. Questions of form to be considered in steel casting are discussed later (page 93).

Malleable cast iron

Malleable cast iron is obtained from white cast iron by annealing treatment. It is made in three qualities known as commercial malleable cast iron, high-grade whiteheart malleable cast iron, and high-grade blackheart malleable cast iron. Its strength figures are higher than those of grey cast iron, and can be taken as approximately the same as those of plain unalloyed machinery steel. Compared with grey cast iron it has the advantage of ductility. Therefore, if fairly large forces are involved and if elongation is called for, then malleable cast iron may be used to advantage if the work entails small intricate castings.

Iron castings with spheroidal graphite

This material also provides considerable ductility and its mechanical properties approach those of mild steel.

Unalloyed steels

Assuming that the designer's careful consideration of a problem has led him to decide on using steel, he will first try to manage with the unalloyed carbon steels. Detailed tables of these machinery steels will be found in any good engineering handbook. Equipped thus with information on applications, methods of working, and properties of the material, he can select a suitable grade.

The information he will find on common sizes available from stock is of the utmost importance for economic reasons. As far as unalloyed steels are concerned, there are no very big differences in price. If stressing is not of a very high order, he will find these steels adequate, since they provide tensile strengths up to 85kg/mm² (54 tons/in²) with satisfactory elongation.

For lightweight construction purposes, screws and bolts can be made in high-tensile alloy steels. Despite their higher price they are still economical to use, because of their smaller dimensions and because of the smaller flanges which result.

Steel sections are standardized in B.S. 4: 1962.

For ordinary structures in sheet and plate such as tanks or air receivers, there is available a range of structural sheet and plate (see Appendix). For steam boiler construction there are special provisions regarding strength of materials. Where necessary, the designer must make a special study of this subject.

The ordinary machinery steels are not suitable for work entailing case hardening and heat treatment. Steels suitable for these purposes are listed in the specification. Where fairly high stressing is encountered, the latter group is suitable for the smaller sizes of forging. For larger forgings, however, it is necessary to resort to alloy steels, since the unalloyed grades do not permit sufficient through-hardening and through-quenching and tempering. The steels having less than 0·2 per cent carbon are used when exceptionally hard surfaces are required.

Alloy steels

If the designer cannot find the properties he wants among unalloyed steels, he can call on a great range of alloy steels. There are over 300 grades, a third of which can be used as structural steels. By the addition of various alloying elements such as nickel, chromium, manganese, tungsten, molybdenum, etc., the various properties of the steels can be significantly improved. Unfortunately, the properties are not usually all improved to the same degree. For example, there are extra-high-strength steels having high Brinell hardness but low elongation and high notch sensitivity. The designer must therefore take great care to choose the grades which are best suited to his requirements. He will be well advised to consider again very thoroughly which properties are the ones which call for his special attention. A graphic representation can give valuable assistance in this task (see fig. 28).

From this example it will be seen that special importance is attached to maximum tensile strength, wear resistance, and weldability. The material selected must satisfy these three requirements without fail. If weldability is not entirely satisfactory, then the requirement is not fulfilled. It is unfortunately not possible to achieve the ideal situation where all the properties required are in fact met by those present in the material. The

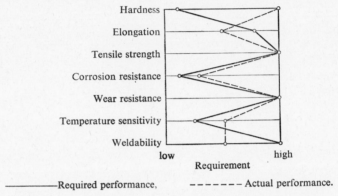

————————Required performance,   – – – – – – Actual performance.

Fig. 28.—Diagram showing utilization of given properties of a material

problem of material selection therefore usually permits only of an approximate solution or a compromise. This means that the designer will often have to decide on a certain grade from among several available steels. The criterion for the final decision is that the steel selected shall fulfil all the vital requirements. Other factors such as machinability and, more particularly, price, also play an important part.

Alloy steels are standardized. To find what he wants quickly, the designer must know the effects of the various alloying elements on the properties of steel. A short guide is therefore given below.

Carbon: With increasing carbon content the tensile strength rises from 30 to 100kg/mm^2 (20–65tons/in²), the elongation drops from 28% to 2%, the hardness increases from 100 to 350 Brinell, and weldability declines.

Silicon: Silicon gives dense castings. With rising silicon content the elastic limit is raised, but forgeability and weldability are reduced.

Manganese: Manganese has the primary effect of increasing wear resistance.

Chromium: Chromium increases the tensile strength, elongation, and hardness. Chrome steels are rust-resistant and highly heat-resistant.

Nickel: The addition of nickel improves tensile strength and elongation. Nickel steels are rust-resistant and highly heat-resistant; they are frequently used in conjunction with chromium to give chrome-nickel steels.

Molybdenum: Molybdenum increases strength at elevated temperatures, makes for better through quenching and tempering, and eliminates temper-brittleness.

Phosphorus: Phosphorus improves strength, but reduces elongation and impact strength.

Tungsten: Tungsten acts to increase hardness.

Copper: Copper improves rust resistance.

Aluminium: Aluminium steels are suitable for nitriding.

By suitably alloying one or more of these substances with iron, certain properties can be brought into special prominence. The following distinctions can be made, for example:

High-strength steels: Cr-, CrNiMo-, Mn- and MnSi-steels.

Highly heat-resistant steels: Cr-steels.

Wear-resistant steels: Mn-, MnMo-steels and CrMoW-steels.

Rust-resistant steels: Cr-, CrNi-, CrMo-steels and the copper-containing steels.

Sufficient information and data on this subject will be found in engineering handbooks.

Non-ferrous metals

The non-ferrous metals of primary importance in mechanical engineering are the following alloys:

> Copper-zinc alloys (brass)
> Copper-tin alloys (bronzes, red brass)
> Aluminium alloys
> Manganese alloys

The brasses offer a series of useful qualities. For fittings and small housings good use can be made of the cast alloys. The special grades can even be used for fairly high-stressed components. The hardness, too, is considerable. A further advantage is the resistance offered to corrosion. Strength is not high at elevated temperatures, however. Temperatures of 200° C should generally not be exceeded.

The wrought alloys are suitable for hot pressing, forging, bending, and cold rolling. For these purposes also the designer has available grades offering good mechanical properties such as high strength, elongation, and hardness. All further details can be obtained from the British Standard Specifications.

Copper-tin alloys.—The bronzes and red brass are widely used in mechanical engineering. These are materials possessing truly favourable properties. Tensile strengths, admittedly, are not high but appear to be suitable for mechanical engineering purposes and for the construction of process plant and fittings. These alloys have good running properties as bearing materials under conditions of high surface loading. They are corrosion-resistant. On the other hand, they are sensitive to temperatures above 200–250° C.

Through the addition of aluminium, bronzes acquire very favourable properties. Mechanical strength and resistance to chemical attack are markedly increased, so that bronzes of this kind can be used for severely stressed parts like gears and worm wheels. The aluminium bronzes are not very suitable for bearings, and certainly not for bearings involving high surface loading.

Lead bronzes.—Lead bronzes are primarily used for heavily loaded bearings and bushes. They are suitable for the heaviest loadings up to 350 atmospheres at high velocities of sliding. The only disadvantage with lead bronze is that it has no running-in capacity and is very sensitive to misalignment. Bearings made of this special alloy must therefore be diamond-bored with the utmost accuracy.

Aluminium alloys.—In a variety of alloyed forms aluminium has had imparted to it very useful properties, with the result that it has become a material with which the designer cannot dispense in modern mechanical engineering. At the present time there are available to the designer aluminium alloys having strength properties equal to those of ordinary structural steel. To this they add good corrosion resistance and fatigue strength. Of quite outstanding importance is the low specific gravity (2·7) which has opened up new fields for the application of aluminium alloys wherever large inertia forces are involved. These alloys can be worked hot and cold, they can be cast, forged, welded, soldered, and bonded with adhesives. British Standards give details of an abundance of cast and wrought alloys. In addition, all the aluminium foundries and rolling mills issue prospectuses full of advice on the properties and applications of the various alloys.

If the designer knows exactly what are the requirements to be met in his problem, then this information will enable him to select the suitable material. He can also choose from a large number of rolled sections, special sections, tubes, and sheet.

Strength values depend on the casting method used, whether die casting or sand casting, and where wrought alloys are concerned, on their condition, whether soft, medium-hard, or hard.

There has also been developed a further range of special aluminium alloys having properties making them suitable for use as pistons, as bearing material, for pressure die-casting work and for applications requiring free-cutting properties.

Magnesium alloys.—The magnesium alloys have a still lower specific gravity (1·8) than the aluminium alloys. They can therefore be usefully employed wherever the maximum amount of weight has to be saved, that is to say in automotive and aircraft engineering. These alloys are divided into cast and wrought varieties. Deformability in the cold condition is limited, but satisfactory work can be produced by forging and extruding at 300° C. They are also readily castable. Magnesium alloys show some strength at high temperatures, but they should not be exposed to temperatures exceeding 200° C. One disadvantage is that they corrode easily. They must therefore be protected by painting, etc.

Wood.—Wood has very attractive properties. Compared with other structural materials, it is easily worked and comparatively cheap, but unfortunately these qualities are offset by others which do not meet with the designer's approval, namely very variable strength and " shrinkage ". Wood is therefore unsuitable for structures which have to be true to size, and even for toleranced components. It can be used with advantage, however, in agriculture for machines and implements, and in scaffolding and staging work and in vehicle building.

Plastics.—The past thirty years have seen the development of a vast number of new plastics which differ widely in their chemical and mechanical behaviour. To make the right choice the designer must know all about the properties of these materials. It is outside the scope of this book to discuss the various plastics available. The designer must acquire the necessary knowledge from a study of maker's catalogues. In addition, detailed tables of these materials are contained in every good engineering handbook.

Plastics can be divided on the basis of their behaviour into curing and non-curing types. Of the latter, Plexiglas has won an important place in the automotive and aircraft industries. The other materials in this group are confined to domestic and art-and-craft applications. The curing plastics by themselves offer only low strength. This can be increased however, by fillers which make these materials suitable for use in mechanical engineering as bearings, gears, screws, piping and housing components for small machines and apparatus. Questions of shape as it affects these materials will be discussed later under " Form Design ".

E. MANUFACTURE

The factors influencing manufacture

After determining the most satisfactory basic design to meet the customer's requirements, the designer's next task is to select the right materials for the various components. We have seen that the desired objective is quickly reached with the aid of an evaluation plan. Determination of the material to be used also decides broadly the manufacturing method to be adopted. For example, if the decision has been taken to use steel owing to the very small quantity of components required and to the good mechanical properties offered, then hammer forging or welding are the manufacturing methods which first suggest themselves. It will be seen, therefore, that the manufacturing method is strongly influenced both by the material used and by the quantity required. The later stages in the processing of the item will be determined by its detailed design, appearance, surface finish, availability of machines, tools and equipment, etc.

For the sake of completeness, we will repeat the list of factors which may influence the manufacturing method. They are:

Form	Jigs, fixtures, and tools
Material	Gauges
Appearance	Fits
Quality of surface finish	Patents
Climatic influences	Quantity required
Available equipment	Cost
Delivery date	Overhaul

At this stage special reference must be made to the factor of manufacturing cost. The economics of manufacture are of such outstanding importance in the designer's activities that they largely determine the course taken in the later stages of a design. The following are the primary considerations of which the beginner should continually remind himself until they have become part of his mental equipment so that he uses them unconsciously at the right time. Avoid the following:

1. Heavy wastage of material	6. Expensive machining operations
2. Re-clamping of work	7. Finishing operations
3. Auxiliary equipment	8. Rejects
4. Special-purpose tools	9. Difficult assembly
5. Change of machine	

It has already been pointed out that fruitful work will only be performed in practice if there is cooperation between the drawing office and the production shops. How else is it possible for the designer to make the necessary allowances for the work load on the machine shop at any given time, and to take existing gauge and tool equipment, etc., into consideration?

The designer is assumed to have a knowledge of production engineering. Information on this is provided as a special study in college courses, so that there is no point in going into it in any detail in a book of this kind. A beginner who finds gaps in his knowledge of production techniques can only be advised to fill them as quickly as possible. In his practical work, too, he will find that he is constantly obliged to learn about new methods, because the new materials continually being developed call for special methods of working.

F. FORM DESIGN

What does the designer have to bear in mind in form design?

For the outsider, the determination of form is the most obvious activity associated with designing. For this reason it is widely held to be the only activity, or the most important one, involved in designing. Even in many of the colleges it was thought that a special subject termed the Theory of Form Design could teach in full the intellectual operations of designing. The entire problem of designing is often looked at only from the point of view of form design. For the same reason, also, it is this aspect which is made the subject of most of the suggestions offered in the way of "wrong" and "right" examples, "hints for the designer" and "guiding principles" for form design. The student designer will have discovered already, however, that determination of form is only a part of the whole design effort.

Form design is decisively affected by many factors:

1. Working principle (basic design)	9. Use of existing products
2. Mechanical stressing	10. Appearance
3. Material	11. Handling and operation
4. Manufacture (cost)	12. Maintenance
5. Considerations of space	13. Overhaul
6. Size	14. Surface condition
7. Weight	15. Fitness for shipment
8. Standard items	16. Power requirement

Before we go on to examine the more important of these influencing factors, it will be necessary to discuss some general points regarding form design.

General points regarding form design

We have seen that all the steps taken in formulating a design must be carried out to a definite plan if the designer wants to avoid the risk of continually running into blind alleys and doing unnecessary unproductive work.

A clear aim is necessary in form design. An experienced designer once said that what was needed was " more designing and less inventing ". And this, indeed, only confirms the requirement stated above, for inventing is widely regarded as an intellectual activity which proceeds intuitively by fits and starts.

Proven designs should never be abandoned without cogent reasons. There are some designers, and they are mostly beginners, who fall into the error of wanting to improve on everything already in existence. It should be borne in mind that other designers, too, have devoted much careful consideration to their designs. Perhaps there is even some special experience embodied in the existing version—experience which one does not oneself yet know about. This is why existing designs should always be carefully studied but only adopted when the intentions and ideas of the other designer are accurately known.

With beginners one notices that they readily become discouraged if they have to change their designs often. These young people should bear in mind that it is a great deal easier to change a design ten times on paper than it is to change it even once in the workshop.

The scales which the designer will use in his drawing work are, of course, laid down for him. For reductions he will use the scales 1:2, 1:4 and 1:8 and for enlargements 2:1, 4:1 and 8:1. He should naturally practise using these scales and should make himself able to visualize the actual dimensions. Despite this it is advisable whenever possible to draw full size.

It is a general practice to draw attention to modified dimensions by underlining. If a number of dimensions have been changed, however, it is better to alter the drawing.

Some designers are very sensitive to any criticism of their work. The right course is to listen calmly to any criticism made, whether expert or not; in this way new ideas will often suggest themselves to the designer.

Every designer should prepare a file specially for his calculations, and this file should be kept with the drawings. It is an important document in the event of complaints and may play a decisive part if legal proceedings are involved.

1. HOW THE BASIC DESIGN INFLUENCES FORM DESIGN

The basic design, as determined by previous analysis, represents only a skeleton which, however, suffices to indicate the fundamental layout. Of course, there is still a long way to go to reach the final form of the desired product, but nevertheless the basic design already incorporates the main idea behind the solution to the problem posed.

It is also very instructive for the beginner to set himself the opposite kind of problem by drawing an existing machine or device in diagrammatic form only. To do this he must learn to pick out the essentials and how to illustrate them.

The designer may find himself having to do this more often than the beginner might suppose. In patent applications, for example, it is the general practice to provide the desired drawing in as diagrammatic a form as possible. A similar situation may arise if the designer somewhere sees a device which is of importance to him. Under these conditions he will only be able to note down the essential facts in a diagrammatic sketch. A representation of this kind is also more readily understandable by the layman and much easier to grasp than the finished product.

2. HOW MECHANICAL LOADING INFLUENCES FORM DESIGN

Every object has extension in space and has certain dimensions and shapes which the designer must first determine from various considerations.

The mechanical loading gives the designer his first datum point in determining the form design and dimensioning of the various parts. Every component of a machine or of a piece of equipment transmits forces which are largely dependent on the layout. The designer must therefore ensure that the force pattern set up is the most favourable one possible, so that the component forces and dimensions can be kept as small as possible. It must also be noted, however, that geometrical form influences strength. This is a fact that cannot be overlooked when dimensioning components.

Most beginners believe that the dimensions of a machine component can be determined right from the start without a drawing. In actual fact, calculation and drawing alternate with each other continuously during design work. An example may illustrate this.

Assume that the hoisting gear for a jib crane has to be calculated and designed. The factors known are the effective load Q, the lifting speed v in m/min, and the height of lift H in metres.

With these particulars the input power to the hoisting unit or the nominal rating of the motor can be calculated provisionally if an estimated figure is used for the efficiency of the gearing. The rope diameter is given by the effective load. The latter determines the smallest allowable drum diameter D. The number of drum revolutions per minute to give the required lifting speed can be found from the expression

$$n_{\text{drum}} = \frac{v}{D\pi} \text{ rev/min}$$

From this stage onwards, points have to be considered which already vitally influence the design. The fact that the electric motor has a considerably higher speed than the drum means that the designer is obliged to provide a gear unit. The necessary reduction ratio i is found from the expression

$$i = \frac{n_{\text{motor}}}{n_{\text{drum}}}$$

This value is governed directly by the speed of the motor.

Now, for a given rating there are wide limits to the speed which may be chosen. If a slow-running motor is selected, the reduction ratio will be smaller and the gear unit simpler and cheaper, but the motor itself will be more expensive. On the other hand, a high-speed motor is cheaper, but the gear unit required will be more complicated and costly. Economic considerations decide the issue in this situation.

But this does not dispose of all the difficulties in the selection of a suitable gear unit. The unit can be built up in various ways. For example, if a ratio of $i = 120$ has been found, then the gearing can be broken down into three pairs of spur gears to give $120 = 4 \times 5 \times 6$, or into a worm reducer and two pairs of spur gears to give $120 = 10 \times 3 \times 4$, or even into a worm reducer and one pair of spur gears to give $120 = 20 \times 6$. These, however, are not the only arrangements possible. How is a decision to be reached in this kind of problem?

The beginner, or the designer who is not sure of himself, starts by looking for an existing design. When he has found a similar set of conditions he takes over the existing design and is reassured because he believes he can counter any objections which may be raised by pointing out that something on similar lines has already been designed and built. He may be right in the majority of instances, but by taking this approach he will never develop into an independently thinking designer.

We need only consider situations where there is no existing example. In this situation the designer must reach a decision from a variety of possible solutions through his own independent efforts. As already pointed out on various occasions, the designer who has not practised this will never be able to work without guidance.

Nor is it immaterial whether the reduction train driven by the motor is broken down into $4 \times 5 \times 6$ or into $6 \times 5 \times 4$, because as the drum is approached so the torques become higher. It may well be, therefore, that a division into $6 \times 5 \times 4$ will give a lighter and thus a cheaper reduction unit. Worm gearing is always more expensive than an equivalent spur-gear unit. It has the advantage, of course, that it can be built to be irreversible if necessary.

The only right way for the designer to reach a decision in a problem of this kind is to work through several assumed layouts and to select the cheapest.

Once the arrangement of the gearing has been decided, it is essential to start planning the hoisting unit so that its dimensions can be fixed. The schematic diagram will look something like fig. 29.

When the basic design has been laid down, the drum should be drawn (fig. 30). A spirally-grooved drum is used to give the rope positive guidance. The rope diameter d and the coiling diameter D are already known by calculation. From the height of lift H the number of turns can be calculated. From this the width B can be calculated after allowance has been made for the rope thickness and for spare turns. An estimated value is used provisionally for the drum-wall thickness.

The shaft cannot be dimensioned until we have a clearer idea of its length. The first question to be asked is whether the driving gear shall be combined with the drum itself or mounted outside the bearing. A simple check shows immediately that the last-mentioned position is the more expensive. We will therefore assemble the driving gear to the drum. Now, how large should be the pitch circle diameter of this gear? It stands to reason that it must be larger than the drum diameter, and also large enough to prevent the worm reducer which drives the pinion from fouling the drum. The next step, therefore, is to find its principal dimension, the module, the worm-wheel diameter, and the worm diameter.

The worm-wheel torque is found from the moment due to the load $QD/2$, allowance being made for the reduction through the spur gearing. With this information, the module for the worm reducer can now be calculated in the usual manner, the necessary assumptions being made with regard to the material for worm and wheel, number of starts, and reduction

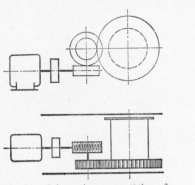

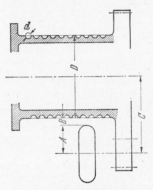

Fig. 29.—Schematic representation of hoisting unit

Fig. 30.—Rope drum with reduction gearing

ratio. After the outside diameter A of the worm unit housing has been determined, and after a small distance B from the drum has been allowed, the centre distance C of the spur gears is reached.

It is now possible to determine provisional radii r_1 and r_2 for the spur gears, and then to ascertain the tooth loading from the moment due to the load $QD/2$. The material for the teeth of the large gear will be selected with due regard to tooth loading and speed of sliding. With these assumptions made, the module, face width, and the exact pitch-circle diameters of the large and small gears can be found.

The drawing now takes on the appearance of fig. 30. But already there arise a whole series of new problems. Should the drum be a fixture on the shaft or revolve on it? What is the best method of attaching the large gear to the drum? The beginner and the detail designer accustomed to copying will now search again eagerly for an existing design on which to base the rest of his own work. If beginners are asked why, for example, they choose to make the drum a fixture on the shaft, the stereotyped answer is nearly always received: because firm X does it that way. Sometimes, however, the beginner notes that firm Y adopts a different method. Then, if he has a critical sense, he starts to ask himself why.

This is the point where a piece of general advice can be given to the student designer, namely that he should try by himself to ascertain the reasons behind the various solutions to practical design problems. After all, we have already seen that differing requirements lead to differing

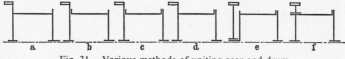

Fig. 31.—Various methods of uniting gear and drum

solutions. To discover the advantages and disadvantages of different solutions to one and the same problem, it is only necessary to work systematically through the various points detailed in connection with the operational and realization tasks (p. 23). If the beginner finds the answer by his own efforts he will feel he has achieved something; if he cannot find the reasons for existing solutions, he can always consult an experienced designer. On no account should the designer take over existing designs without appreciating their advantages and disadvantages.

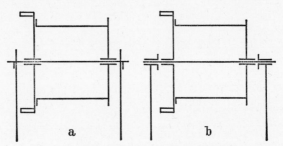

Fig. 32.—-Different bearing arrangements for drum

Returning to our example, the first problem to be solved is how to connect gear and drum. The independent designer does not rely thoughtlessly on some existing design, but thinks out all the various layouts. Schematic diagrams are sufficient for this purpose (fig. 31).

The unpractised designer can use the familiar evaluation plan method to assist him in reaching his decision; it must take into account the factors of pattern cost, core support, core removal, and machining cost. The experienced designer will pick out versions *b*, *c*, and *d* and will probably decide on *b*.

The only outstanding problem now is to decide on the drum bearing arrangements. There are of course two possible layouts (fig. 32). The solution of this problem does not call for long experience. It can be solved by the knowledge possessed by a beginner. The right decision is so obvious that no evaluation plan is needed. For the sake of clearer understanding, the advantages and disadvantages of versions *a* and *b* are compared in Table 2. The advantages of version *a* are so great that the disadvantage implicit in the more involved lubrication method is willingly accepted.

TABLE 2

Factor considered	Version *a*	Version *b*
Stressing of shaft	Static, no notch effect	Alternating, notch effect through keyway
Bearings	Bearings do not need any specially rigid mountings	Rigid bearing mountings needed
	Simple, cheap bearing arrangement	Massive and expensive bearing arrangement
	Somewhat restricted access to bearings	Easy access to bearings

We will decide, therefore, for version *a*. With this decision taken, the drum gear can be planned far enough to give the overall width (fig. 33). The bearing loads *G* and *H* are found by calculating from the two end positions E and F. With this information, the loading of the shaft and the pressures it exerts on its supports can be found. Consequently, after the material has been chosen, the shaft size for bending can be worked out. We now draw the shaft, giving it the diameter we have just found, and make an estimate of the specific bearing loading. For relatively light loading no bushes are required. The drum with its bearings can now be finally planned. All that remains is to check the assumed drum-wall thickness for torsion and the gear for strength. For attaching the gear to the drum there is no need to use expensive fitted rings; instead fitted bolts or split tubular dowels with ordinary bolts can be used.

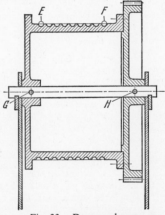

Fig. 33.—Drum and gear

It is now time to leave discussion of this problem. The purpose of the example quoted was to show the student designer that in design work the calculations cannot be carried out in isolation from the form design; rather is it on the progress of the drawing work that the constant development of new principles for further calculation depends. Alongside this there are, in form design, many questions to be answered concerning choice of material, manufacturing method, etc. We will discuss these further in later sections.

The factors affecting form design which the designer must take into account with regard to mechanical stressing can be summarized in the following rules.

Rules

1. Try to secure a simple force transmission arrangement which avoids regions of high stress.

The situation is ideal when only tensile or compression stresses exist in the component. This can often be achieved. If, for example, the problem is to mount a bearing bracket via a flange on to two channel members (fig. 34), then it would be unsatisfactory

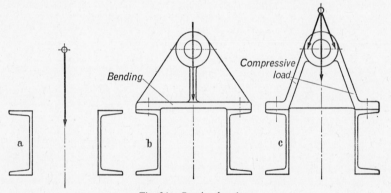

Fig. 34.—Bearing bracket

to choose version *b* because here the bottom flange is subjected to bending stress. This contrasts with version *c* where the ribs have to carry compressive forces only. Where cast iron is concerned this factor plays an important part as we shall see later.

2. The above findings can be summarized in the following rule: Place the material so that it follows the same direction as the lines of force.

This rule, of course, expresses a natural law of which there are many examples in the morphology of plants and animals.

3. Wherever possible use shapes which can be expected to give satisfactory, that is to say low, stress conditions, such as cylinders, cones, and spherical surfaces.

Such surfaces are also easy to manufacture. This can be regarded as a law of life, for wherever strength is required nature avoids flat surfaces and designs exclusively on the basis of circular forms.

The beginner tends to give preference to flat forms. This can be seen from the following example (fig. 35). If the task presented is to develop a housing and mounting flange for a pair of bevel gears, then the majority will choose shapes resembling version *b*, although version *c* is simpler, cheaper, stiffer, more pleasing in appearance, and better from the casting point of view.

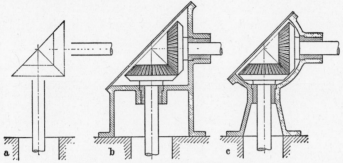

Fig. 35.—Housing for a pair of bevel gears

4. Determine the external and internal forces present.

This, indeed, is a requirement to be met before starting on any design task. A catch question familiar to engineers provides a good illustration of the difference between external and internal forces. If a skier carries his skis on his shoulder as shown in fig. 36, then they will exert a force *B'* which will be greater than the weight of

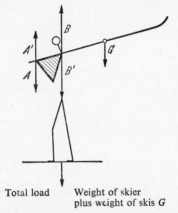

Total load   Weight of skier
                  plus weight of skis *G*

Fig. 36.—Ground force exerted by a skier

skis on his shoulder. Does this mean that the skier is now heavier by the load *B'* exerted on his shoulder? The answer of course is no. He can only be heavier by the weight *G* of the skis since forces *A'* and *B'* act as internal forces in the closed triangle formed by shoulder, arm, and ends of skis.

5. Consider whether any additional inertia forces, elastic deformations, and impact or shock loads will arise.

Such inertia forces are always undesirable, and their elimination often causes the designer great difficulty. When resilient members are interposed, these may even give rise to resonance effects which can easily lead to fracture. Examples which are probably familiar to the reader are the flexural oscillation of steam turbine shafts brought about by an eccentric location of the overall centre of gravity of the wheels, and the racking vibration of early-type electric locomotives with side-rod drive. The reason for this was resilience in the transmission mechanism, and the impulsive nature of the bearing forces. The effect of inertia forces is something which has to be kept under observation where vehicles are concerned. An example of this is the car shown in fig. 37.

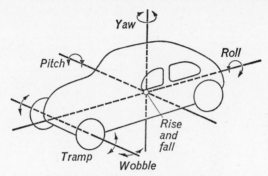

Fig. 37.—Inertia forces affecting a vehicle

6. Always design a component from the inside outwards.

This is a basic law which can be confirmed everywhere in nature; it applies also to architecture. Even if the outline is laid down, the architect will arrange the layout from the inside, and not until this has been done will he plan the façade. The situation in mechanical engineering is quite similar. Here, admittedly, the dimensions relating to volume may be laid down. If, for example, the width dimension is very restricted, then possibilities must be provided for allowing the part to develop in length. Every engineering product demands a certain minimum volume. The impossible cannot be realized. Development, however, always takes place from inwards outwards, although with the chance of deviation in any direction. We shall find examples of this later when discussing the influence of spatial conditions on form design.

7. To begin with, calculate and dimension approximately according to classical theory of the strength of materials.

Eye alone can never be developed to the stage where one can manage in all situations without calculation; especially is this true when new forms are concerned. Not until he has made an approximate calculation will the designer gain a feeling of confidence.

8. Find where the stress-raisers and maximum stresses are, and then:

9. Ask how the maximum stresses can be reduced. This subject is discussed further on page 194.

10. The final calculation and dimensioning are then carried out having regard to the fatigue strength and the calculated stress.

Exercise problems

The student designer must know the laws of statics, kinematics, and dynamics thoroughly. For the solution of the familiar examples set in colleges, etc., there is so much literature available that the student can always find guidance. If, however, during his design work, he comes upon an unfamiliar problem he will be faced by serious difficulties, because examples will usually be lacking and he will be obliged to determine what laws are involved and how the calculation is to be carried out. For practice purposes, therefore, two small problems are given below.

Problem 8

The lazy-tongs principle is used to effect the parallel guidance of a circular saw (fig. 38). Assume that at point A a force *K* acts vertically downwards. Determine the forces in the joints B, C, D, E, F and in the guide G. Also determine the nature of the stressing in the links.

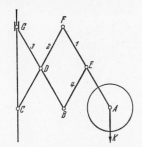

Fig. 38.—Arrangement of a parallel-
guided pendulum saw

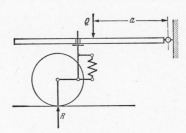

Fig. 39.—Arrangement of a single-
axle trailer

Problem 9

For a single-wheel trailer for a passenger car the payload *Q* and its distance from the pivoting point are known. All the important dimensions needed for further calculation are given in fig. 39.

Find all the external forces acting on the bearings and joints of the wheel-fork and links.

3. INFLUENCE OF MATERIAL ON FORM DESIGN

The material used has an important influence on form. This is true in all cases no matter whether the object is an engineering component, a wood-carving, or a moulded-plastic product. There was a time, of course, when failure to recognize this principle led to cast-iron machine parts being given a style alien to the material and borrowed from Doric or Gothic architecture.

The beginner is often confused by the literature of the subject which provides many examples and much information dealing with the relationship between material and form. It is possible, however, to reduce the question of how the material influences form design to the following simple statement.

When the properties of a material are known, *all that is necessary is to design so that the most satisfactory characteristics are utilized to the full and the least satisfactory avoided as far as possible.*

The properties of materials can always be divided into three groups—physical, chemical, and technological.

It is under these headings that the principal engineering materials capable of being shaped by non-cutting methods, that is to say by casting, forging, pressworking, and moulding, will be examined.

4. HOW THE PRODUCTION METHOD INFLUENCES FORM DESIGN

Every production method leaves its own unmistakable imprint on the outward appearance of an engineering component. Who, for example, could not tell by appearance alone that a casting and a welded structure are quite different from each other? Practically any required shape can be produced in cast iron. And this, indeed, was partly responsible for the fact that in the early days of engineering, as mentioned above, there was such complete misunderstanding of function, material, and manufacturing method that bearing pedestals were designed on the lines of Gothic windows. Several decades of development were needed before it was realized that the treatment is determined by the properties of the material, and that the method of treatment leaves its own special mark on the outward form.

The designer must therefore be familiar with all methods of treatment and must bear them continually in mind while at work.

Linked inseparably with the production aspect is the question of cost. The time is long past when everyone was satisfied if the designer turned out a component which was capable of being manufactured and which did its job. Nowadays, faced with fierce competition and the constant demand for greater economy, the designer is forced to aim at the lowest possible cost consistent with a design which is sound from the production viewpoint.

(a) Form design of grey iron castings

The material of paramount importance for general engineering purposes is cast iron. The principles of designing in cast iron will therefore be discussed in some detail.

I. Physical properties

Hardness.—Ordinary grey cast iron has a fairly high hardness figure (about 200 Brinell). For special applications such as crusher jaws, drawing dies, or punches, a very hard grade (chilled cast iron) is available to the designer.

Tensile and compression strength.—Owing to the importance of cast iron, an extract from B.S. 1452:1961 is given in Appendix D (p. 256).

It will be seen from the Appendix that the strength of grey cast iron varies with thickness. Apart from the carbon and silicon content, the graphite separation also is affected by the rate of cooling. The latter, however, depends on the thickness of the casting. This means that sections of different thickness in one and the same casting will differ in strength owing to variations in the crystalline structure.

It follows from this alone that castings should be designed for maximum uniformity of thickness throughout. There are, however, other arguments to support this principle, as we shall see in due course.

The strength figures of the grades listed in Appendix D are lower than those for steel. A purely pearlitic iron with finely distributed graphite may, however, give strength figures equalling or surpassing those of ordinary machinery steel, namely 30–55 kg/mm^2 (20–35 tons/in^2), and may even combine strength of this order with a capacity for elongation which of course is lacking in ordinary grey cast iron.

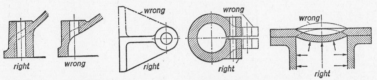

Fig. 40.—Avoidance of tensile and bending stresses

There is a striking difference between the compressive strength and the tensile strength of cast iron, the ratio being about 4:1. Iron castings should therefore be loaded mainly in compression. Although this condition cannot always be achieved, it is usually possible to design so that tensile and bending forces are eliminated or at any rate minimized (fig. 40).

Strength in bending.—It is a characteristic of cast iron that its strength in bending is substantially greater than its strength in tension. The ratio

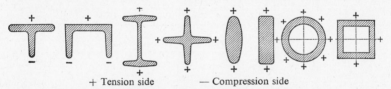

+ Tension side   — Compression side

Fig. 41.—Correct choice of tension side for various
cross-sectional shapes

is in fact about 2:1. The reason for this is that, under conditions of bending, the neutral fibre shifts towards the compression side, so that the tensile forces become smaller. Despite this, however, the designer will take care to ensure that parts subjected to bending stresses are given the most satisfactory possible cross-sectional shapes; these are shown in fig. 41.

It is a well-known fact that a curved section is stronger than a flat section. When designing for casting, therefore, the aim should be to make as much use as possible of cylindrical, conical, and spherical surfaces.

The discovery that the strength in bending of cast iron depends on the cross-sectional shape (fig. 42) dates from Bach. This matter of shape is something which the designer must take into account.

The beginner usually has trouble in deciding what values to use for tensile and transverse stress. And, certainly, his hesitation is justified to

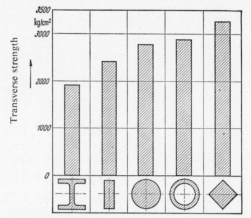

Fig. 42.—Relationship between transverse strength and cross-sectional shape for grey cast iron

some extent, for only if the stress distribution in the workpiece is known with some certainty is it permissible to use the higher-ranging strength values.

Strength in torsion and shear.—On this subject, unfortunately, there is little experimental information available. Strength in torsion can be taken

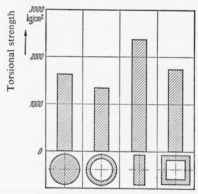

Fig. 43.—Relationship between torsional strength and cross-sectional shape for grey cast iron

as approximately the same as strength in bending. In actual fact, however, the torsional-strength figures also vary with cross-sectional shape (fig. 43).

It would appear from the diagrams that an annular cross-section gives the lowest torsional strength. It should not be concluded, however, that this cross-section is the least suitable one for torsional loading. If cross-sections of equal area (and therefore of equal weight) are evaluated for their polar moment of inertia, it is found that the annular cross-section gives the highest value, and, consequently, the most satisfactory loading conditions.

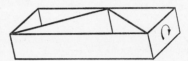

Fig. 44.—Torsionally strong box-shaped member

The forms mainly used for parts stressed in torsion are the cylinder, the cone, and the hollow box-shaped member stiffened by diagonal ribbing (fig. 44) and these have long been employed for machine-tool applications.

Elasticity.—The modulus of elasticity E has a wide range of values and depends on the stress applied. In fact E ranges from 490,000 to 1,130,000 kg/cm^2 (3100–7200 tons/in^2), and may therefore be about 50 per cent lower than that for steel. This means that for a given stress cast iron deflects more than steel. More metal must therefore be used if rigidity equal to that of steel is to be obtained.

Elongation at fracture.—For ordinary grey iron this is so small that for all practical purposes it can be put at zero. Cast iron is therefore not suitable for any part subjected to impact loads if the part concerned must respond to loading by deforming and not by suddenly fracturing.

Heat resistance.—A phenomenon peculiar to cast iron is " growth ". Up to about 350°C the strength of cast iron is little affected by temperature. Above 400°C, however, the combined carbon decomposes into ferrite and graphite. At the same time a dangerous increase in volume takes place, and this sets up stresses in the casting which superimpose themselves on existing stresses and so lead to the formation of cracks at the critical cross-sections. The strength also naturally decreases to a marked extent above 400°C. For this reason steam-turbine casings exposed to steam temperatures above 400°C are made of cast steel, which does not suffer from this dangerous characteristic.

Wear resistance.—The wear resistance of cast iron also depends on the crystalline structure. With decreasing pearlite the structure becomes wear-resisting, especially in the presence of secondary cementite and ledeburite. This type of cast iron is suitable for parts subjected to heavy wear, such as road rollers, crusher jaws, or drawing dies.

As far as sliding properties are concerned, cast iron behaves in much the same way as bronze or red brass. It is therefore used for bearings, pistons, slide valves, etc. When used for bearings the surface can be ground and polished. It is very sensitive to misalignment, and it is therefore essential to ensure that the housing bore is perfectly lined up with the shaft axis. The figures for permissible loading per unit area and for peripheral speed are lower than for bronze, red brass or white metal.

II. Chemical properties

A structural material is required to be resistant primarily to water, damp air, acids, and bases. In general it can be stated that cast iron is not resistant to rusting in the presence of water and atmospheric moisture. On the other hand there is a grade which is resistant to acids and alkalis and which has useful applications in the chemical industry. Corrosion effects can, however, be avoided by suitable surface treatment, including paint and other protective coatings, cladding and electroplating, galvanizing, tin-plating, and lead-coating.

III. Technological properties

In the foundry use is made of the properties metals have of becoming fluid and thereby castable at suitably high temperatures. The molten metal is then poured into moulds having cavities corresponding in shape and size to the component being cast. After the metal has cooled and solidified, the casting is removed from the mould, fettled, and finish-machined on a variety of machine tools.

The designer must, of course, have a thorough knowledge of the technology of making castings, so that he can make allowance in his designs for the many foundry requirements which have to be met.

It is not the function of a book of this kind, however, to impart knowledge on this subject, which has its rightful place, along with suitable practical exercises, in lectures on the technology of production. All that will be done here is to discuss the way in which technological and production factors influence form design.

All the rules regarding form design can be traced back to:

1. Properties inherent in the metal cast: (*a*) castability, (*b*) segregation effects, (*c*) contraction, (*d*) shrinkage cavity formation.
2. Foundry requirements.

The principal rules to be observed in form design will now be discussed under these two headings.

Factors affecting castability.—To ensure proper flow of metal and filling up of the mould, it is not essential to round corners and slope surfaces. This means that there would be no difficulty even in casting shapes like fig. 45*a* if other factors did not play a part.

During casting, there is evolution of gas (bubbles) which must be allowed to make their way upwards. This is achieved by sloping the walls or additional ribs (fig. 45*b*, *d*, and *f*). The means to be provided for taking the air away are the concern of the foundryman.

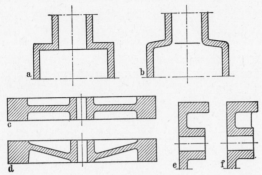

Fig. 45.—Factors affecting the form design of iron castings

Factors affecting segregation effects.—During casting, splashes of metal are often formed, and at low casting temperatures these result in the solidification of sulphur-iron alloys in globular form; they are detrimental to the quality of the casting. These effects are only of significance to the foundryman.

Factors affecting shrinkage-cavity formation.—The cold casting has a smaller volume than the molten one. The total change of volume is termed

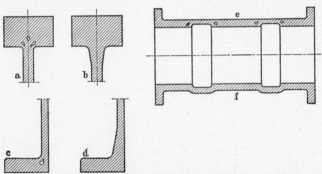

Fig. 46.—How the factor of shrinkage-cavity formation affects
the form design of cast iron

" contraction ". The undesired consequences of contraction are shrinkage-cavity formation and the setting up of stresses. These cavities result from the fact that the outside of the casting and the thinner parts are the first to cool and solidify. The contraction effect causes the molten cast iron to be drawn in. If its attempt to do this is prevented or restricted, voids called shrinkage cavities will be formed (fig. 46*a*, *c*, and *e*). For this reason, gradual changes of cross-section and uniform thicknesses are to be recommended (fig. 46*b*, *d*, and *f*). The foundryman for his part has various methods of preventing unequal cooling; these include premature uncovering, the use of chills to facilitate the removal of heat, the provision of feeders, etc.

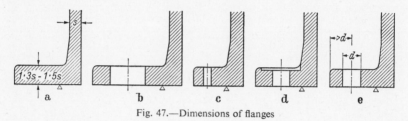

Fig. 47.—Dimensions of flanges

An ample silicon content acts to reduce contraction. During the design stage, the designer for his part must take steps to prevent the formation of shrinkage cavities. He does this by designing for the greatest possible uniformity of wall thickness, by arranging for changes in section to be gradual, and by avoiding concentrations of material.

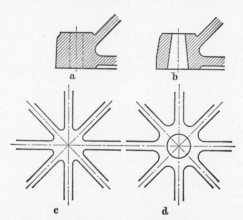

Fig. 48.—Avoiding concentrations of material
(*a*) and (*c*) wrong, (*b*) and (*d*) right

While on this subject, reference will be made to an error constantly committed by beginners. The thickness of a flange is made greater than the wall thickness, not equal to it (fig. 47*a*); this is done both for reasons of strength and also to allow for the machining of the contacting surfaces. At the same time of course a gradual change of section from the wall to the flange must be adopted (standardized). The bolt thickness, too, must stand in a reasonable relationship to the flange thickness, unlike the versions shown in *b* and *c*. Bolts should be set as close to the wall of the casting as possible, but not in such a manner that the fillet is cut into as in version *d*. The distance from the centre line of the bolt to the edge of the flange is usually made too large by the beginner, although this does not contribute anything to the strength of the flange. This distance only needs to be made somewhat larger than *d* (diagram *e*).

Avoiding concentrations of materials.—The risk of shrinkage cavities exists wherever there are concentrations of material. When designing the form of castings, therefore, the designer must ensure that this is avoided (fig. 48*b* and *d*).

Avoiding distortion and casting stresses.—A further advantage is obtained by insisting on uniform wall thickness. This advantage lies in the

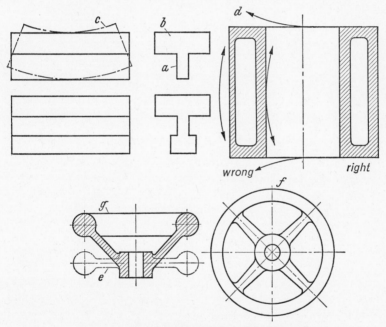

Fig. 49.—Avoiding distortion and casting stresses

avoidance of undesirable casting stresses. These dangerous stresses arise through inequality in the rate of cooling of the different sections of a casting owing to differences in thickness, and this in turn results in unequal contraction. For example, if rib *a* in fig. 49 cools first and solidifies, flange *b* will still be at a higher temperature owing to its greater thickness. As the cooling of the flange proceeds, rib *a* will prevent it from contracting. The tensile stress set up will deform the whole piece in the way indicated in *c*. Similar examples are shown in *d* and *e*. Contraction can be avoided by suitable arrangement of the parting lines as in *f*, or by placing the spokes at an angle as in *g*. The designer should not rely on the expedients already mentioned for obtaining a uniform rate of cooling, but should design to avoid the risk of locked-up stresses.

Factors affecting removal of pattern.—To enable the pattern to be lifted out of the mould without any risk of tearing away the moulding sand, it

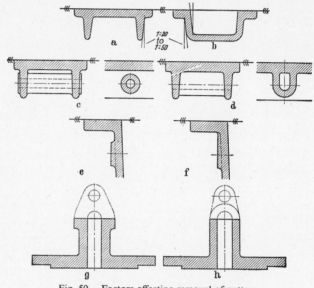

Fig. 50.—Factors affecting removal of pattern

(*c*), (*e*), (*g*), wrong   (*d*), (*f*), (*h*) right

is necessary to give draft to the walls in the direction of lifting (fig. 50*a* and *b*).

Removal from the mould may also be hampered by projecting details (fig. 50*c*, *e*, and *g*). The remedy is to adopt shapes such as *d*, *f*, and *h*.

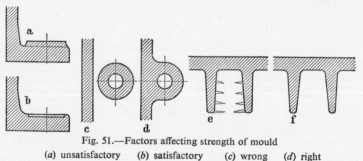

Fig. 51.—Factors affecting strength of mould

(*a*) unsatisfactory   (*b*) satisfactory   (*c*) wrong   (*d*) right

Factors affecting strength of mould.—In a sand mould thin sections are to be avoided, since they will break off easily or become perforated while casting. Designs of the kind shown in fig. 51a and c should therefore be avoided, and the layouts shown in b and d adopted instead. If ribs are spaced very close together it may happen that the pattern will tear the mould when being lifted out (fig. 51e). A more pronounced draft on the walls of the rib will eliminate this danger (see version f).

Rounding of edges.—Beginners are often not clear about the rounding of edges. Indeed, it is " minor details " like this that show whether the

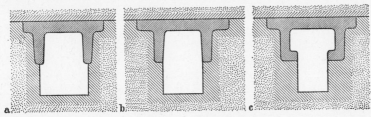

Fig. 52.—Rounding of edges

student designer is already familiar with the way in which shape is governed by the method of moulding (fig. 52a, b, and c). Machined surfaces always form a sharp edge where they meet a rough cast surface. These edges can of course easily be chamfered. Often, however, there is no urgent reason why they should be rounded. In addition, the walls of a casting

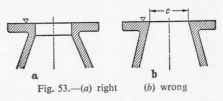

Fig. 53.—(a) right   (b) wrong

should never run obliquely (fig. 53b) into the working surface, but should approach it as perpendicularly as possible (version a), since otherwise dimension c will alter in the course of machining.

Correct disposition of draft.—The arrangement of sloping surfaces depends on how the parting lines are positioned (fig. 54b, c, and d).

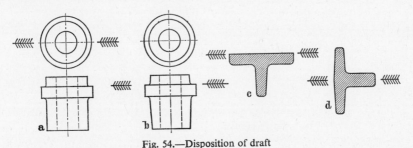

Fig. 54.—Disposition of draft

Providing for simple core shapes.—Castings containing hollow features require what are known as " cores ". These make the moulding work more expensive. Therefore, if the designer is obliged to use cores, he should endeavour to give them the simplest possible shapes, using straightforward

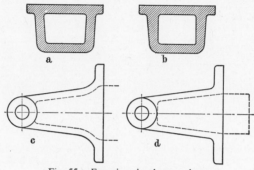

Fig. 55.—Ensuring simple core shapes
(*a*), (*c*) expensive   (*b*), (*d*) cheap

surfaces which are easy to produce. Examples of this are given in fig. 55*b* and *d*, in contrast with *a* and *c*.

Ensuring proper core support.—The cores must be capable of being properly supported so that they do not become misplaced during casting.

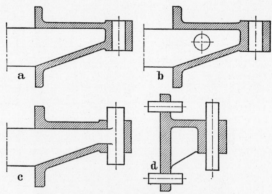

Fig. 56.—Ensuring satisfactory core support
(*a*) wrong   (*b*) unsatisfactory

The arrangement shown in fig. 56*a*, for example, is not recommendable because the core is left self-supporting. For an item of this kind further support must be provided on the lines of either *b* or *c*. If several cores are needed, they are best positioned in the same parting plane (fig. 56*d*).

Avoiding cores.—By appropriate form-design measures the designer often succeeds in avoiding the use of cores. A very familiar example of this is the partly enclosed cavity shown in fig. 57c. Moulds for this kind of work always require an expensive core, and if ribs are provided these must be given openings for core support purposes.

The modified shape shown in *d* can be moulded in a pair of boxes without the use of a core. Types *a* and *b* show examples of cast-in bolt

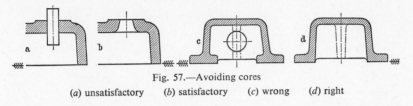

Fig. 57.—Avoiding cores

(*a*) unsatisfactory   (*b*) satisfactory   (*c*) wrong   (*d*) right

holes. It is difficult to say with any exactness what is the smallest diameter of bolt hole which can still be economically cast in. It is not likely, however, to be less than 20 mm. It is possible to manage without a core if the bolt hole is given the tapered form shown in *b*.

Choice of vertical or horizontal casting.—It is often difficult to decide whether a casting shall be made by the vertical or by the horizontal method.

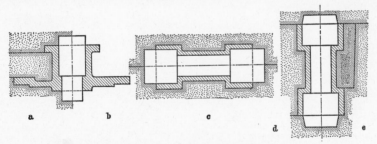

Fig. 58.—Choice of vertical or horizontal casting

(*a*) and (*d*) moulded vertically: 3 boxes, 1 core
(*b*) and (*c*) moulded horizontally: 2 boxes, 1 core
(*e*) moulded vertically: 2 boxes, 2 cores

In doubtful situations it is best for the designer to discuss the question with the works. By appropriate form-design methods, however, the designer can certainly simplify and cheapen the moulding operation quite considerably. For example, the valve *d* (fig. 58) requires two boxes and two cores if cast vertically, but only two boxes and one core if cast horizontally (version *c*). The demand for denseness of structure in a casting of course also decisively influences the moulding method adopted. The cylinder cover *a* requires three boxes and a core if cast vertically, but only two boxes and a core if cast horizontally in the modified form shown at *b*.

Securing dense structure in castings.—In any casting, whether poured vertically or horizontally, the denser structure is at the bottom. If, for example, a vertically cast cylinder (fig. 59a) is required to have a densely cast top flange, this is secured by placing a feeder on the flange. When the casting has solidified the feeder head is cut off. If surface A of the cover is required to have a dense structure it will be necessary to mould this part in such a way that surface A is at the bottom (version c).

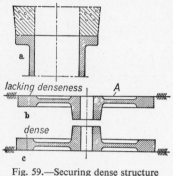

Fig. 59.—Securing dense structure in castings

Ribbed or hollow castings.—By their outward shape, castings can be divided into two types, namely ribbed and hollow castings. Hollow castings have a cleaner appearance and are aesthetically preferable to ribbed castings. They are dearer, however, owing to the preparation of the cores required. With ribbed castings the ribs may be placed so unsatisfactorily that involved and costly moulding is needed (fig. 60b and c) compared with the properly designed versions a and d.

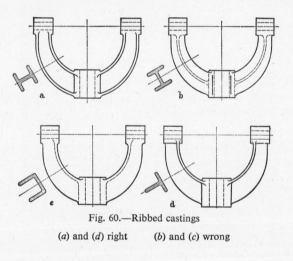

Fig. 60.—Ribbed castings

(a) and (d) right   (b) and (c) wrong

Pattern or template.—For castings of circular shape, the template method of moulding is often most suitable. This method makes expensive wooden patterns unnecessary. Shapes produced by the template method also have the advantage of being perfectly circular in contrast with wooden patterns which may distort. The decision between pattern moulding and template moulding is governed by purely economic considerations. With template moulding, of course, the patternmaker's wages and timber costs are eliminated, but moulder's wages are a bigger item. Template moulding is therefore suitable only for one-off or small-lot production, whilst for long-run and mass-production purposes only pattern

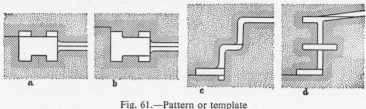

Fig. 61.—Pattern or template

(*a*) Pattern   (*b*) Template   (*c*) Pattern or template   (*d*) Template

moulding is practicable. The question of the minimum number of items for which template moulding is still economical is one which must be decided individually for each application.

All of the shapes shown in fig. 61*a*, *b*, *c*, and *d* can be made by the template method. The advantage of the template method is brought out particularly clearly in shape *d*. To make this shape by pattern moulding would require three boxes and two cores, whereas when using the template method only two boxes are needed.

Provision of machining allowance.—The beginner may well overlook the fact that allowances must be provided on the pattern for machining pur-

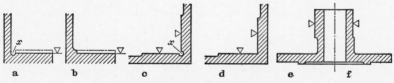

Fig. 62.—Providing machining allowance

(*a*), (*c*), (*e*) wrong   (*b*), (*d*), (*f*) right

poses. This often leads to the use of impracticable forms (fig. 62*a*, *c* and *e*). Parts as thin as those indicated at the corners *x* cannot be moulded in sand of the usual composition. The remedy is to use the forms shown in *b* and *d*. Type *e* would require a loose piece, whereas type *f* is easy to lift from the mould.

It should be noted at this stage that the bosses so widely used are best

4

avoided altogether (fig. 63). They add to the cost and may become misplaced on the casting to the detriment of its appearance. The form shown
in *b* should therefore be used.

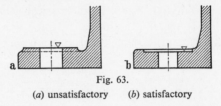

Fig. 63.

(*a*) unsatisfactory   (*b*) satisfactory

Avoidance of rambling shapes.—For strength reasons alone, the designer
should endeavour to give his castings compact shapes. Castings with wide-
spreading rambling shapes are therefore to be avoided. Examples of this

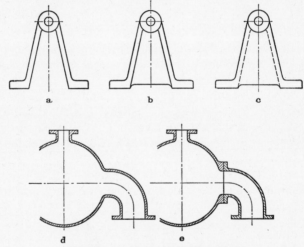

Fig. 64.—Avoidance of rambling shapes

can be seen in fig. 64*a* and *d*. The remedy is to choose a suitable shape
(version *b* or *c*) or to divide the casting into smaller units (version *e*).

Before he starts on the task of designing a casting, the beginner should
at least call to mind the major principles determining form, and act accordingly.

Rules

1. Choose the most satisfactory form from the point of view of strength.
2. Avoid cast iron where impact and shock loading are involved.
3. Avoid cast iron at temperatures above 300°C.
4. Design the casting for maximum uniformity of wall thickness.
5. Avoid sharp corners and concentrations of material.

6. Ensure that changes of section are gradual.
7. Provide draft where necessary.
8. Provide radiusing where necessary.
9. Avoid partly closed forms, attachments, and closed cavities.
10. Take care that cores are straightforward and properly supported.

Exercise problems

Problem 10

A wall-mounted bracket is required in cast iron. The dimensions to be observed are given in fig. 65. The journal pin is a fixture in the hub. The forces acting on the two projecting pivots are equal and constant, but opposed.

Make suggestions for a suitable design.

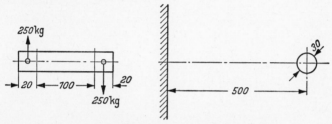

Fig. 65.—Specification for a wall-mounted bracket

Problem 11

What can be criticized in the casting shown in fig. 66? Two leading dimensions are given as an indication of size, and the loading is shown by the resultants of the belt pulls.

Make suggestions for modifying the bracket.

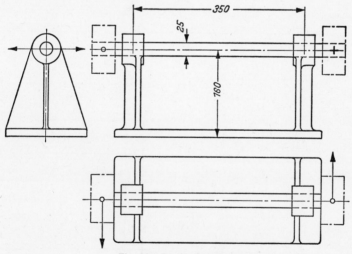

Fig. 66.—Bearing bracket

Problem 12

A return pulley for a belt is arranged to revolve on a shaft (fig. 67). The shaft is held a fixture in the two supports A and B.

The intention is to make the two shaft supports a fixture with tube C via a fork-shaped casting, care being taken to comply with the dimensions indicated. Which of the well-known cross-sectional forms is most suitable for the fork arms?

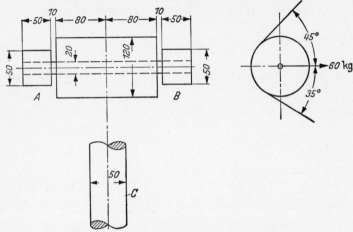

Fig. 67.—Details for a fork-shaped bearing bracket

Problem 13

A pump set, consisting of pump and electric motor, is to be mounted on a cast-iron bedplate. Design the latter with due regard to proper form and strength. Fig. 68 gives the measurements of the mounting surfaces of the motor and pump, as well as the height of the bedplate. Assume the torque transmitted by the motor to be 4 m-kg.

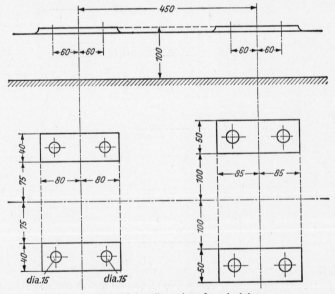

Fig. 68.—Fixing dimensions for a bedplate

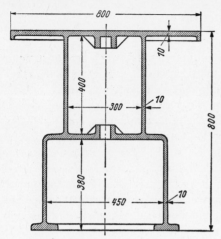

Problem 14

Develop the design of the casting shown diagrammatically in fig. 69 so that it is right from the moulding and casting viewpoints. The dimensions are given as a guide to the size of the casting. Modify the casting where necessary so that its manufacture will not present any difficulty as a moulding or casting proposition.

Fig. 69.—Table in grey cast iron

Problem 15

How should the twin lever be moulded to avoid the use of cores? (fig. 70).
Design the cross-sections of the lever arms and indicate the position of the parting plane.

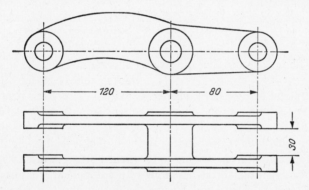

Fig. 70.—Twin lever

(b) Form design of steel castings

Cast steel has greater strength than cast iron. It is therefore always used instead of cast iron when loading conditions are relatively severe and it is desired to retain the advantages of casting.

In the form design of steel castings it is of course equally necessary to take into account the properties of the material. We will therefore review the properties of cast steel again briefly and design accordingly.

I. Physical properties

Steel castings are made in four ordinary grades which are adequate for most applications in general mechanical engineering. For special duty there are available to the designer special grades and alloy cast steel.

Hardness.—The extreme hardness of cast steel makes it equally as suitable as cast iron for crusher jaws, edge mill components, etc.

Tensile and compressive strength.—Cast steel has strength properties similar to those of ordinary machinery steel. Since its structure is not as dense as in rolled or forged products, it is necessary to use somewhat lower strength values. The tensile strength, however, is two to four times greater than that of cast iron. The compressive strength is approximately the same as the tensile strength. There is therefore no need, as there is with cast iron,

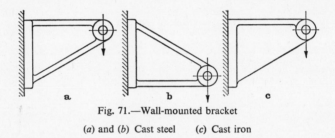

Fig. 71.—Wall-mounted bracket

(*a*) and (*b*) Cast steel   (*c*) Cast iron

to worry about avoiding tensile stress. This means that the shapes used are different and much freer. The bracket shown in fig. 71*a* and *b* could be made in cast steel, whereas if made in cast iron, type *c* would be the most suitable with a view to the tensile and compressive loading conditions.

Bending strength.—The bending strength of cast steel is about twice as high as that of cast iron. Owing to the greater strength of cast steel compared with cast iron, the re-designing of a component from cast iron to cast steel results in smaller dimensions, and, therefore, lighter weight.

Strength in torsion.—As far as torsional strength is concerned, the difference between cast iron and cast steel is not as pronounced as with tensile and transverse loading, but nonetheless a figure 50 per cent higher than for cast iron can be used.

The strength properties of cast steel can be very considerably improved by the addition of copper, manganese, nickel, etc.

Elasticity.—The modulus of elasticity of cast steel is of the order of twice that of cast iron, so that cast steel is better suited than cast iron for structures requiring greater stiffness. Cast iron, however, has better damp-ing capacity. By suitable form design it is possible to give steel castings fatigue resistance also—a property which is required, for example, in vehicle components.

Elongation at fracture.—Elongation at fracture for cast steel is 12 to 20 per cent. Cast steel is therefore very suitable for parts which when severely loaded shall only take on an elastic or permanent deformation and shall not suddenly fracture.

Heat resistance.—The causes of " growth " were explained earlier in connection with cast iron. This phenomenon does not occur in cast steel because there is no free graphite present. Cast steel can therefore be used at temperatures above 400° C.

Wear resistance.—Cast steel is given greater toughness and wear resistance when alloyed with manganese. It can then be used whenever parts are required to offer considerable resistance to wear as in components for mechanical shovels and crushing machines.

Cast steel is unsuitable as a bearing material, but is used for backing shells for bearings which are then lined with white metal.

II. Chemical properties

There is a special grade of cast steel which is rustless. Its resistance to acids and bases is still not widely enough known.

Surface treatment is possible as with cast iron (see page 80).

III. Technological properties

The principles of designing in cast iron can be applied also to cast steel in all points where the properties of the two materials do not differ; these points are therefore primarily the ones concerning proper design from the pattern-making and moulding viewpoint. With cast steel, however, special care is necessary where the properties differ.

Cast steel has a higher melting point (1450° C) than cast iron. It flows sluggishly, tends to form bubbles more readily, and has twice as much contraction. The foundryman's task is therefore more difficult than with cast iron.

All the form design aspects of these properties therefore require more careful treatment compared with cast iron. The critical factors are those which make the design correct from the point of view of casting, accuracy of size, and provision for cleaning.

Since the risk of shrinkage-cavity formation is greater with cast steel than with cast iron, the designer must take even greater care to ensure that the castings he designs have the greatest possible uniformity of wall thickness.

The main thing is to avoid constrictions by using suitably chosen

radiusing (fig. 72). The use of feeders is extremely important wherever
concentrations of metal occur. The designer must therefore see that these
are provided for in his design, and that they can also be easily removed

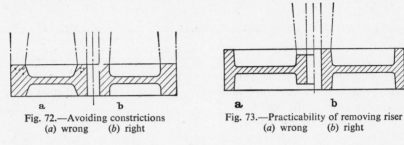

Fig. 72.—Avoiding constrictions
(*a*) wrong   (*b*) right

Fig. 73.—Practicability of removing riser
(*a*) wrong   (*b*) right

after casting (fig. 73*b*). Steel casting calls for risers on a much larger scale
than is needed for cast iron if shrinkage cavities are to be avoided and
bubbles and slag carried away. Risers often amount to 50 to 100 per cent
of the weight of the workpiece.

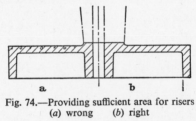

Fig. 74.—Providing sufficient area for risers
(*a*) wrong   (*b*) right

Surfaces intended to take risers must be made sufficiently large (fig.
74*b*) otherwise there is a risk of shrinkage cavities forming as at *a*.

Surfaces required to be dense when cast will be placed at the bottom
as with cast iron. If this is impracticable, the alternative is to provide a
slag bead which can be trimmed off after casting.

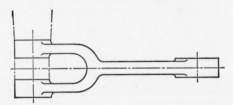

Fig. 75.—Ensuring denseness of casting at bearing bosses

To obtain a dense cast structure in the fork shown in fig. 75 it is
necessary to cast solid to allow bubbles to escape upward so that shrinkage
cavities cannot form.

The check circle method has established itself as a means of detecting concentrations of metal.

A concentration of metal always occurs at the point where two walls of equal thickness come together (fig. 76a). Such concentration can be

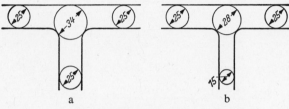

a   b

Fig. 76.—Avoiding concentrations of metal

(a) wrong   (b) right

reduced, however, if one of the walls is made thinner (version b). This is why, for steel castings, ribs are always made thinner than the wall in the ratio

$$\frac{\text{rib thickness}}{\text{wall thickness}} = 0\cdot6 \text{ to } 0\cdot8$$

and why fillets having a radius of one-third to one-quarter of the wall thickness are used.

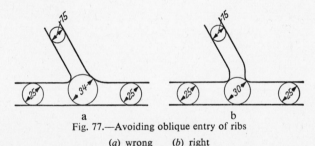

a   b

Fig. 77.—Avoiding oblique entry of ribs

(a) wrong   (b) right

A rib which enters a wall at an angle gives a bigger concentration of metal than one which enters perpendicularly (fig. 77). The former should therefore be avoided wherever possible.

The fact that cast steel contracts twice as much as cast iron raises the risk of hot cracks occurring while in the mould. For a length of 500 mm the contraction is as much as 10 mm. If the mould will not yield to this extent there will be a risk of cracks occurring at the critical points (fig. 78a). One method of preventing this is to reinforce the weakest point by what are known as *contraction ribs* which can be trimmed off afterwards if necessary; or the critical points may be strengthened by ribs (fig. 79*b*).

Owing to the large amount of contraction, the size variation also is larger with cast steel than with cast iron. The designer must therefore aim

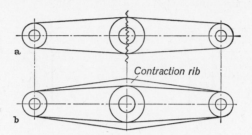

Fig. 78.—Contraction rib to prevent cracking

to secure maximum uniformity in the contraction which occurs (by using uniform wall thicknesses, avoiding concentrations of metal, etc.). Castings can often be straightened while in the red-hot condition.

Fettling also is more difficult with cast steel than with cast iron. The large amount of contraction causes cores to be held more tightly in the

Fig. 79.—Boss reinforcement to prevent cracking

casting so that they are more difficult to remove. The designer must allow for this by making the cores readily accessible and easily removable.

To sum up, it may be stated that all the guiding principles set out for the form design of cast iron also apply to cast steel. In view of the different properties of cast steel, however, special attention should be given to:

Rules

1. Take great care to avoid concentrations of metal.
2. Use risers generously.
3. Allow sufficient area for risers to stand on.
4. Provide for convenient removal of risers.
5. Avoid constrictions by making proper use of radiusing.
6. Aim at maximum uniformity of wall thicknesses.
7. Always make ribs thinner than walls.

8. Prevent hot cracks by providing contraction ribs.
9. Ensure proper core support.
10. Ensure easy removability of cores.

Exercise problems

Problem 16

The cross head of a slewing crane is often made as a forging (fig. 80).

If required in quantity, however, there might well be a need to produce it in cast steel.

The external forces acting on the journals are indicated in fig. 80.

Try to design the cross head as a steel casting making the thrust bearing a rolling bearing.

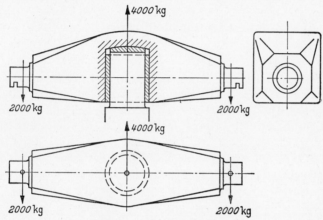

Fig. 80.—Cross head (forged)

Problem 17

A link having the measurements indicated in fig. 81 is loaded as shown. It makes only a rocking movement 30° either side of the mid-position. The two outer eyes are to be designed as bearings. A fixed pin passes through the central hole.

Design the link as an iron casting and a steel casting.

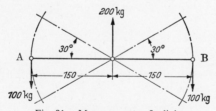

Fig. 81.—Measurements of a link

(c) Form design of malleable iron castings

Malleable cast iron is used for parts which owing to their small size and intricate shape cannot be made as steel castings or forgings, and for which cast iron does not appear suitable owing to the comparatively high strength and elongation required. The question whether to make a part as a malleable iron casting or a forging is largely determined by the quantity required, and must be decided individually for the part in question on a costing basis.

Malleable cast iron is therefore used only for fairly small items ranging from 0·5 to 2 kg, such as spanners, fittings, chains, wheels, levers, shaped parts, parts of agricultural machines, etc. Parts weighing up to 30 kg can, however, also be made in malleable cast iron; examples of this are to be found in automotive engineering (differential housings). The principles to be followed in designing are again dependent on the special properties of the material.

Malleable cast iron is made in two qualities.

I. Physical properties

Hardness.—Malleable cast iron is hard and can therefore be used for grinding and crushing wheels, wearing surfaces, sand-blasting nozzles, etc.

Strength.—The strength figures yielded by malleable cast iron are higher than those for cast iron, but lower than the values for cast steel. Parts for which malleable cast iron is used are not likely to be severely loaded. It is the property of toughness which plays the more important role. The strength figures can be reckoned to be the same as those for ordinary machinery steel. Sizes, therefore, are also similar to those of parts made of the latter.

Heat resistance.—The tensile strength falls off considerably from 400°C upwards. There is, however, also a heat-resistant grade which can be used for melting pots.

Wear resistance.—To impart high wear resistance, blackheart malleable cast iron can be hardened. Malleable cast iron is not suitable as a bearing material, not even for bearing shells. Bearing bores in malleable cast iron must therefore be bushed to make plain bearings.

II. Chemical properties

A corrosion-resistant grade is available. Surface treatment is possible with malleable cast iron. It can be lead-coated and tin-plated.

III. Technological properties

Malleable cast iron is made by pouring white cast iron of suitable composition into moulds. A subsequent annealing process gives one of the following products, depending on the atmosphere used:

(a) Whiteheart malleable cast iron if the atmosphere is oxidizing.
(b) Blackheart malleable cast iron if the atmosphere is neutral.

The different methods used in making malleable iron castings compared with ordinary iron castings give rise to special design rules which must be borne in mind alongside the rules applying to cast iron generally.

In the making of whiteheart malleable cast iron the malleableizing action is only effective down to depths of a few millimetres. For this reason thick sections should be avoided, and instead wall thicknesses of 8 to 15 mm should be used, in other words shallow, T-shaped, I-shaped, U-shaped and ┿-shaped sections.

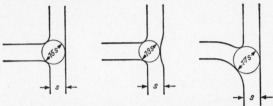

Fig. 82.—Checking concentrations of metal

In blackheart malleable cast iron the temper carbon is distributed uniformly across the whole cross-section if the annealing has been properly carried out. Blackheart has a more uniform structure than whiteheart. Cross-sectional shapes chosen may be as for cast iron, and wall thicknesses may range up to 30 mm.

The subsequent annealing treatment may cause distortion, cracks, and fractures. This danger is particularly great with whiteheart malleable cast iron. When one realizes that contraction may amount to 2 per cent it is

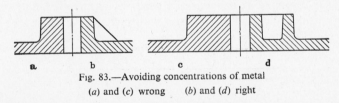

a   b   c   d

Fig. 83.—Avoiding concentrations of metal

(*a*) and (*c*) wrong   (*b*) and (*d*) right

clear that the risk of shrinkage cavities is greater than with cast iron. The designer must therefore aim to avoid anything which may lead to a concentration of metal, such as sharp corners, abrupt changes of section, and casting holes solid.

The check circle can again be used for testing radiusing with a view to avoiding concentrations of metal at meeting points (fig. 82).

Where white-fracture malleable cast iron is concerned, concentrations of metal are exceptionally dangerous owing to shrinkage-cavity formation. In addition, the inner core remains hard owing to deficient decarburization. This can be remedied by using ribs (fig. 83) or recesses (fig. 83*d*). An alternative method would be to make version *a* in blackheart malleable cast iron.

When made of cast iron, forked members are often cast solid so that gases can escape upwards and allow a dense structure to form. With

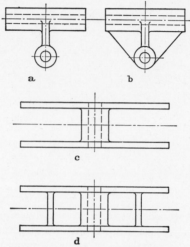

Fig. 84.—Avoidance of rambling shapes
(*a*) and (*c*) rambling   (*b*) and (*d*) right

whiteheart malleable cast iron, however, this practice must be avoided owing to the greater tendency to shrinkage-cavity formation.

Widely overhanging portions of a workpiece generally tend to distort during casting. With malleable cast iron this danger is even greater owing

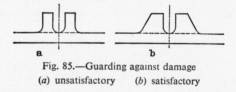

Fig. 85.—Guarding against damage
(*a*) unsatisfactory   (*b*) satisfactory

to the larger amount of contraction and the subsequent annealing process. Malleable iron castings can, of course, be straightened again, but nevertheless the aim should be to adopt a form which will render corrective treatment unnecessary. The designer can best guard against distortion by providing webs or ribs (fig. 84).

Malleable iron castings are usually fettled twice. Excessively thin features should be avoided, otherwise there is a risk that they will be damaged (fig. 85*b*).

In the foundry malleable cast iron is treated in the same way as ordinary cast iron. Consequently the same rules of form design are to be observed in like manner. The subsequent annealing treatment and the large amount of contraction, however, oblige the designer to take special precautions. These are summarized below.

Rules

1. With blackheart malleable cast iron, all that matters for practical purposes is the casting process. Observe the same principles as for ordinary cast iron.
2. With whiteheart malleable cast iron there are further factors to be borne in mind in addition to the familiar foundry requirements; these factors entail:
   (*a*) the use of uniform wall thicknesses (8 to 15mm),
   (*b*) the avoidance of concentrations of metal,
   (*c*) the use of appropriate blending radii at meeting points,
   (*d*) the use of ribs and recesses instead of large concentrations of metal,
   (*e*) the avoidance of cores wherever possible.

* * * * *

Non-ferrous alloys.—The principal non-ferrous alloys used in mechanical engineering are as follows:

1. Copper-zinc alloys (brass and tombac)
2. Copper-tin alloys (bronze and red brass)
3. Aluminium-bronzes
4. Lead-bronzes
5. Aluminium alloys

The first three varieties are used for comparatively small items like gears, worms, housings, and fittings. The strength properties of these materials can be found from any good engineering handbook. The copper-zinc and copper-tin alloys do not offer high strength; their strength properties are about equal to those of grey cast iron. Their hardness is very much lower. On the other hand, they show considerable elongation.

Higher strength is exhibited by the aluminium-bronzes. These possess tensile strength ranging from 35 to 45kg/mm^2 (22–28 tons/in^2)—figures which correspond to those of ordinary machinery steel. In Brinell hardness, too, these alloys are not greatly inferior to steel. All three types of alloy display excellent sliding properties, and they are therefore used for bearing shells under conditions involving very severe loading and speeds of sliding.

The castable grades of these alloys possess good properties from the foundry point of view, and they are therefore subject to the same considerations as apply to the form design of cast iron (see pages 80 to 90).

The aluminium alloys call for special experience on the part of the foundryman, owing to the possibility of the cast being affected by slagging brought about by the ready oxidation of aluminium.

The lead-bronzes, too, are difficult to handle in the foundry owing to their tendency to demix, with the result that stratification may occur. Their tensile strength is very low. Their use as die-casting material is therefore restricted to smallish castings as used in light engineering. As bearing materials, the lead-bronzes can be used under extremely heavy loading up to 5000 lb/in^2, but the bearing shells must be diamond-bored to the exact tolerance specified, since this type of alloy possesses no running-in properties.

(d) Form design of aluminium castings

In recent years the light metals have found their way into all branches of mechanical engineering on a steadily increasing scale. The reason for this is to be found chiefly in their low specific gravity, which is about two-thirds less than that of any of the heavy metals. The light metals can be machined more readily and cheaply, and at higher cutting speeds than other metals. To this must be added the possibilities of plastic cold and hot forming, as well as the many different methods of fastening which can be employed, such as welding, soldering, bonding with adhesives, riveting, screwing, and folding.

It is true that light metals cost considerably more than other materials, but this fact is partly offset by the lighter weight.

The strength properties of aluminium alloys are extremely favourable and equal those of ordinary steel.

An advantage which should not be under-estimated is the high resistance to chemical attack offered by many of the alloys, so that it is easy to find a material suited to a given purpose.

Furthermore, various types of surface protection can be used, and these still further extend the range of application to a considerable degree.

As a result of these many advantages, light metals have been put to use in the construction of internal-combustion engines, motor vehicles, and aircraft, principally because of their light weight and the advantages which go with it.

Their resistance to chemical attack makes the aluminium alloys specially suited for use in the chemical industry and in the field of domestic equipment.

A vast range of sheet, strip, bars, and tubes in a great diversity of profiles is available to enable light metals to be used on the widest scale. The aluminium alloys are listed in all the well-known engineering handbooks.

The distinctive properties of the light metals give rise to many differences in design treatment.

I. Physical properties

Specific gravity.—The most outstanding property of the light metals is their low specific gravity compared with the heavy metals (2·7). It is this property which opens up uses for the aluminium alloys wherever an exceptionally lightweight mode of construction is specified. A more detailed treatment of the design factors to be considered will be found in the section on lightweight construction on page 186.

Hardness.—All the light metals give very low hardness figures. They are therefore unsuitable for designs exposed to impact, shock, or surface damage. Damage in the form of nicks and scratches impairs the strength of these materials which are very notch-sensitive.

Strength.—There are some aluminium alloys, and they are primarily wrought alloys, which exhibit strength figures similar to those yielded by

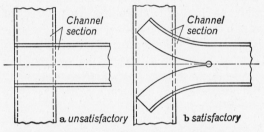

Fig. 86.—Designing for favourable force distribution

ordinary machinery steel. Cast alloys suitable for highly stressed parts are, however, also available to the designer.

The bearing strength is low, and certainly lower than that offered by the heavy metals. When deciding the shape of aluminium parts, therefore, the designer must keep the surface loading lower than for the heavy metals. Areas under compression must be made 25–30 per cent larger than in steel. For this reason, key-type fastenings, which usually entail heavy loading per unit area, cannot be recommended. The load-carrying capacity of regions exposed to surface loading can be increased by steel inserts which can be cast in or pressed in.

The notch sensitivity of the light metals is greater than that of steel. This calls for special precautions in the form of the design. The low hardness in itself demands care in the machining and working of these metals if damage to the surface is to be avoided. The designer must arrange for a steady and gradual transfer of forces and must avoid abrupt transitions (fig. 86b).

Large-radius fillets ensure a satisfactory stress pattern. The radii used in bending and folding operations must also be chosen with care. Sharp notches are definitely to be avoided.

Elongation.—The light metals do not display brittle behaviour like cast iron, but show a fairly large amount of elongation. The aluminium alloys can therefore be used with advantage when deformation rather than sudden fracture is called for, as in automotive and aircraft engineering.

Elasticity.—The modulus of elasticity is considerably lower than that for steel; in fact it is about one-third. Items in aluminium alloys therefore exhibit very great elastic distortion. If they are to be given greater stiffness, their moment of inertia must be increased. This may lead to large cross-sections being given a hollow form, or to the provision of stiffening by means of ribbing. At force-transfer points, special care must be exercised to prevent forcible deformation.

Heat resistance.—" Growth " does not occur because there is no graphite in the structure. On the other hand, all the strength values are extremely sensitive to elevated temperatures. In fact, even at temperatures of 100°C upwards, it is advisable not to use light metals.

Thermal expansion.—Under the same conditions, the thermal expansion of the light metals is about twice as great as that of steel. The designer who uses dissimilar metals must make special allowance for this property, so that no additional stresses will be introduced as a result of the different rates of expansion.

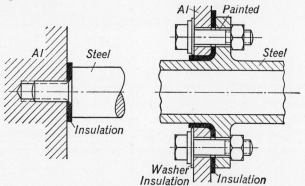

Fig. 87.—Insulation when using dissimilar metals

At this point reference must be made to a property of the aluminium alloys which plays an important role in mechanical engineering. When light metals are assembled to heavy metals, and particularly to copper alloys, electrolytic corrosion cells are set up at the points of contact when moisture is present, and these cells result in the destruction of the aluminium.

To prevent this, the designer must insulate the dissimilar metals from each other by using a suitable spacer material. For ordinary situations, painting with bitumen or lacquer is sufficient. Red lead cannot be used. Rubber and synthetic plastics are excellent insulants (fig. 87).

Screws and bolts must of course be protected also. This can be done by galvanizing or cadmium-plating. The covering must always be applied in such a manner that no current path can form.

Wear resistance.—The aluminium alloys offer only a small amount of resistance to wear, and are therefore unsuitable for applications involving severe wear. This deficiency can, however, be remedied to some extent by suitable surface treatment, including, for example, hard chromium plating or anodizing, and by providing sliding surfaces of steel.

II. Chemical properties

In general, aluminium and its alloys display good corrosion resistance. It is usually possible to find a suitable alloy which is immune to a specific chemical attack. By suitable form-design measures, the weather resistance of these metals can be still further improved, for example through the avoidance of water pockets and places for the accumulation of liquid residues, and through the provision of sloping surfaces and satisfactory radii to help water to run off. For the same reason, sharp corners and edges are to be avoided.

Where necessary, surface protection can easily be given, for example by painting with lacquer, spraying, cladding, or electroplating.

III. Technological properties

Aluminium can be sand cast, gravity die cast, pressure die cast, and centrifugally cast. As far as foundry practice is concerned, the aluminium alloys are more difficult to handle than cast iron. Molten aluminium has a very strong tendency to oxidation. The unfortunate thing about this is that instead of the slag being lighter and rising to the surface, it remains in the melt if not properly treated and leads to inclusions of slag and dross. These in turn impair the strength considerably and result in a lack of denseness in the castings produced. In addition, there is a risk of aluminium castings picking up gas, depending on the casting temperature used.

These are dangers which foundry specialists have to bear in mind and counteract by taking suitable precautions, e.g. by proper planning of the casting system, control of casting temperature, or use of fluxes.

The designer, too, should know about these difficulties because he will then be in a position to design accordingly and to cooperate with the foundry to contribute to the production of satisfactory castings.

As far as the making of patterns for aluminium castings is concerned, the factors to be considered are generally speaking the same as for cast iron. This does not mean to say, however, that a pattern used for an iron casting is equally suitable for casting the same item in aluminium or an aluminium alloy. Quite apart from the greater contraction of the aluminium alloys (1 to 1·4 per cent) there is also the fact that their strength properties differ from those of cast iron. Points where forces are applied or transferred need to be designed more carefully than when using heavy metals. When flange-type fastenings are used, the number of bolts should be made larger than for cast iron or steel, in order to secure better disposition of the force pattern. Washer surfaces also must be made larger, owing to the smaller loading permitted per unit area. It will be seen therefore that patterns intended for iron castings cannot be taken over as they stand.

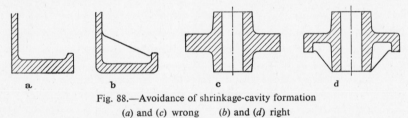

a   b   c   d

Fig. 88.—Avoidance of shrinkage-cavity formation

(a) and (c) wrong   (b) and (d) right

We will now examine the principal factors to be considered when designing aluminium castings.

Patterns for aluminium castings must be clean and smooth, since any rough spot will be reproduced and may give rise to notch effects.

Mould-filling capacity or fluidity, that is to say the capacity to fill the mould accurately, is very high—like cast iron provided that the right casting technique is used.

The risk of shrinkage cavities exists for the same reasons as apply to cast iron. Sinkage holes may also occur. These are sunken areas on the surface of the casting caused mainly by suction effects as the metal solidifies. Wet sand giving rise to steam and consequent pressure on the surface of the casting may also cause superficial cavities of this kind.

The internal-shrinkage cavities occur at points where there are large concentrations of metal and they are brought about, as with cast iron, by contraction and suction (fig. 88). Shrinkage-cavity formation can be lessened by using reduced wall thicknesses and reinforcing by means of ribs (examples b and d).

Aluminium castings may also be affected by miniature shrinkage cavities distributed throughout the cross-section and formed during the solidification phase.

In the course of the earlier remarks on iron castings it was pointed out that the ideal condition would be for the casting to cool uniformly to bring about simultaneous solidification. Unfortunately there is no casting process capable of achieving this condition. This means that during the cooling phase some parts of an aluminium casting will be semi-fluid while others are already solid. In the semi-fluid condition the metal lacks strength and stretch, so that cracks may be caused through contraction at the boundaries with parts already solid. The magnesium alloys are particularly liable to do this.

Contraction cracks and stresses can be avoided by appropriate form design. The designer must see to it that wall thicknesses are as uniform as possible and that abrupt transitions are avoided (fig. 89).

By attention to points of moulding technique, such as using chill plates,

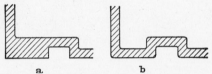

a.　　　　　　b

Fig. 89.—Avoiding contraction cracks and high-stress regions

(*a*) wrong　　　(*b*) right

designing the mould properly, providing risers, and keeping the mould loose, the foundryman can counteract the formation of heat cracks and stresses.

After passing through the zone of hot shortness, that is to say after solidification has taken place, the casting contracts by the characteristic amount of 1–1·4 per cent. As a result of this, stresses and cracks are likely to be set up in the casting if it is prevented from contracting. The countermeasures to be adopted by the designer are the same as for the form designing of iron castings.

It is the foundry's responsibility to plan the casting system. The designer for his part, however, must see to it that large risers can be placed in position and also removed again easily (fig. 73).

Even for cast iron it is better to replace the horizontal faces of a casting by sloping ones. For the purpose of aluminium casting this is all the more necessary, since slag and dross may be trapped at these points and lead to the formation of shrinkage cavities and stresses. Thick-walled features offer less strength than thin-walled. No workpiece should be cast with a thickness under 4 mm, however, although small items in aluminium alloys can be cast down to 2 mm.

Ribbing is used on aluminium castings for the simple reason that,

owing to the smaller wall thicknesses employed, stiffening must be provided in order to obtain greater strength. The points to be borne in mind
by the designer in this connection are as follows. Unduly high ribs (fig.

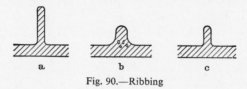

Fig. 90.—Ribbing

90a) are liable to set up stress peaks. Excessively thick ribs tend to cause
shrinkage cavities (example b); for optimum results, comparatively low
and thick ribs should be used (example c).

Fig. 91.—Stiffening corrugations

Instead of the conventional ribs, corrugations as shown in fig. 91b can
be very usefully employed for stiffening purposes. These are resilient and
do not tend to set up stresses and cracks.

The designer using aluminium and its alloys must choose a form yielding the best possible stress pattern. Failing this, the extra stresses set up

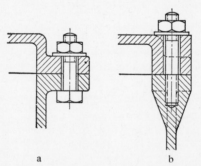

Fig. 92.—Avoiding unsatisfactory force pattern

(a) unsatisfactory   (b) satisfactory

will cause severe deformation owing to the low modulus of elasticity of
aluminium and its alloys (fig. 92).

For this reason, heavy bolts are best replaced by a number of lighter
ones. Screw-type fastenings which often have to be dismantled should be
in the form of through bolts rather than cap screws.

As far as design of the cores is concerned, the same casting and mould-ing principles apply as in the manufacture of grey-iron castings. Cores must be of simple shape. They should be easy to introduce into the mould and must be properly located. The latter point deserves special attention, since any displacement will have more serious results than with

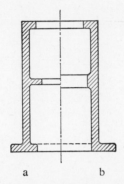

a   b

Fig. 93.—Avoiding waisted cores

(*a*) unsatisfactory   (*b*) satisfactory

cast iron, owing to the reduced wall thickness of aluminium castings. Partly-closed shapes and waisted cores which may easily break down should also be avoided (fig. 93).

The practice of casting in metallic-core supports is one to avoid.

To supplement the form-design principles discussed earlier in con-nection with grey-iron castings, the points to be specially noted when designing aluminium castings are listed below.

Rules

1. Make wall thicknesses as uniform as possible and use small machining allow-ances.
2. Keep wall thicknesses small and stiffen where necessary.
3. Avoid concentrations of metal by providing ribs or recesses.
4. Avoid abrupt transitions.
5. Provide for the placing of risers.
6. Replace horizontal surfaces by sloping ones.
7. Make sure that the stress pattern is as favourable as possible.
8. Make sure that cores are properly located.
9. Avoid using waisted cores.

Exercise problem

Problem 18

The design for a welded cover (fig. 94) is to be produced as an aluminium casting offering the same strength. Sketch a suitable design.

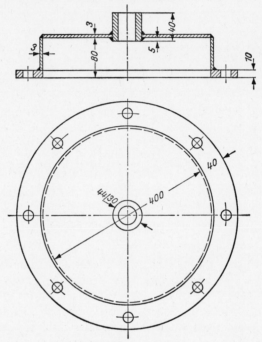

Fig. 94.—Design for cover

(e) Form design of pressure die castings

As the name implies, this process introduces the metal into the mould under pressure; it is a casting method suitable only for mass-production runs. For runs of less than 1000 items it is seldom economical to use pressure die casting. Only comparatively small parts are made by this method. Details of typical weights for the various alloys concerned are as follows:

Lead alloys: 1 kg max.   Aluminium alloys: 10 kg max.
Tin alloys: 0·5 kg max.   Magnesium alloys: 8 kg max.
Zinc alloys: 20 kg max.   Copper alloys: 5 kg max.

Comparison of the specific gravities reveals figures of much the same order for zinc, aluminium, and magnesium alloys. Pressure die casting does not have the same importance in mechanical engineering as do sand casting and gravity die casting. It is of great importance in light engineering, however, and in this field runs of several tens of thousands or hundreds of thousands of items are common.

Compared with other production methods in which shape is imparted without cutting, pressure die casting offers substantial advantages, as follows:

Direct conversion of raw material into finished product.
Economy in use of material owing to small amount of waste.
Speed of the production process.
High accuracy of finished parts.
Interchangeability of castings produced.
Clean finish and high surface quality.
Hardly any finishing off required.

Earlier, when cast iron was under consideration, it was seen that the principles of form design can only be stated if the properties of the material and the characteristics of the casting process are taken into account. Before we proceed any further it must be made clear that the proper design of pressure die castings is more difficult than that of grey-iron castings.

A whole range of alloys is available to the designer for pressure-die-casting purposes. The alloys used are as follows:

Lead pressure-die-casting alloys
Tin pressure-die-casting alloys
Zinc pressure-die-casting alloys
Aluminium pressure-die-casting alloys
Magnesium pressure-die-casting alloys
Copper pressure-die-casting alloys

Further information on this subject will be found in engineering handbooks. With the necessary particulars known, the designer can make his choice to suit the needs arising in service, the shape required, and the properties of the material.

The strength factor.—Reasons connected with casting technique and with the low strength values of the alloys used make it impracticable to produce highly stressed parts as die castings when substantial wall thicknesses are required. For static loading, and assuming wall thicknesses up to 4mm, the following figures can be taken as the permitted range of tensile stress for the alloys concerned:

Zinc alloys: $1-2\,kg/mm^2$ $(0{\cdot}6-1{\cdot}2\,tons/in^2)$
Aluminium alloys: $3{\cdot}5\,kg/mm^2$ $(2{\cdot}2\,tons/in^2)$ max.
Magnesium alloys: $2{\cdot}5\,kg/mm^2$ $(1{\cdot}5\,tons/in^2)$ max.

For parts in compression the position is more satisfactory. Allowable compressive stress can be taken as 50 per cent higher than the figures for tensile stress.

For surfaces bearing one on the other, the following figures can be taken for the permissible loading per unit area:

Zinc alloys: approx. 1·5–3 kg/mm^2 (1–2 tons/in^2)
Aluminium alloys: approx. 5–6 kg/mm^2 (3–4 tons/in^2)
Magnesium alloys: approx. 4–5 kg/mm^2 (2·5–3 tons/in^2)

Transverse strength and torsional strength are about equal to tensile strength, whereas the strength in shear is only half the tensile figure. When repeated alternations of stress are experienced, the figures are naturally much lower and amount to about one-quarter to one-fifth of the tensile strength. Consequently, notches in particular are to be avoided.

Casting-in of inserts.—A practice which immediately suggests itself as a way of increasing the strength of pressure-die-casting alloys is to introduce into the casting parts which are made of high-strength materials. If this is done, however, the following points must be watched:

1. The pressure-die-casting alloy must shrink firmly on to the insert.
2. The cross-section of the insert should not be so large that the surrounding die-cast metal splits away (fig. 95a).

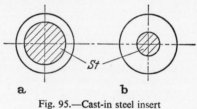

Fig. 95.—Cast-in steel insert

(*a*) wrong   (*b*) right

3. The insert must be capable of being properly anchored and located in the die (fig. 96).

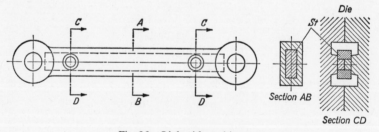

Fig. 96.—Link with steel insert

Threads can of course also be cast in as an integral feature. For highly stressed internal threads, however, it is better to cast in bushes or nuts (see examples in fig. 97). For better retention in the casting and to prevent rotation, grooves, notches, threads, multiple flats, etc. are used (fig. 98).

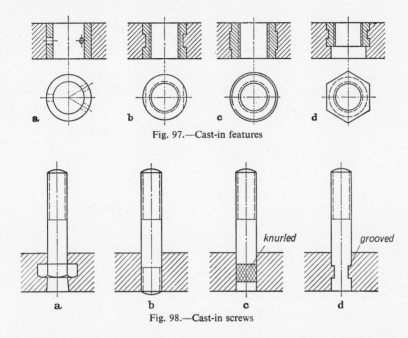

Fig. 97.—Cast-in features

Fig. 98.—Cast-in screws

Factors affecting castability.—In pressure die casting the filling of the dies does not take place as easily as the filling of the mould in grey-iron casting. The often intricate configuration of the dies, the various changes of direction at corners, and the thinness of the walls are serious obstacles

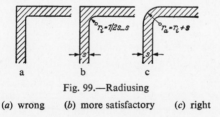

Fig. 99.—Radiusing

(*a*) wrong   (*b*) more satisfactory   (*c*) right

to the filling of the die cavity. This means that the metal must be forced into the die cavity at high speed and fused together completely before cooling occurs. It is important therefore that radiusing should not be made too small (fig. 99). The sizes of the radii used will depend on wall thicknesses and the type of alloy employed.

Uniform wall thicknesses.—Any lack of uniformity in wall thicknesses (fig. 100) will promote shrinkage cavities and set up stresses as well which usually lead to cracks. The thinner section solidifies more rapidly than the thicker, and contracts in the arrowed direction. Owing to the lack of yield in the steel dies, this process gives rise to tensile stresses which are much more dangerous than those occurring in sand casting and which may

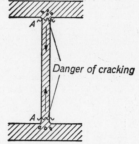

Fig. 100.—Consequences of non-uniform wall thickness

easily lead to cracking at A. Therefore, for pressure die casting also, the designer should make it a rule to use wall thicknesses of maximum uniformity.

If certain parts of a pressure die casting are required to have increased wall thicknesses for reasons of extra strength, the remedy is to provide ribbing (figs. 83, 88, and 89).

Avoiding concentrations of metal.—If radii are made too large, the result will be a concentration of metal, and the consequences of this will

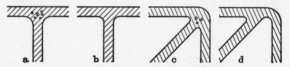

Fig. 101.—Avoiding concentrations of metal

be similar to those noted in connection with grey-iron castings, namely shrinkage-cavity formation (fig. 101). The radius used should not exceed the wall thickness.

The choice of the right wall thickness for a pressure die casting calls for more detailed consideration than when dealing with sand castings. If the wall thickness is made too large, there will be a risk of shrinkage cavities and of the surface sinking through contraction. The factors on which choice of wall thickness chiefly depend are as follows:

1. The type of alloy used, since there are both low-melting-point and high-melting-point alloys for pressure die casting.
2. The length of the path taken by the molten metal.
3. The area involved.

The last-mentioned point is important because, if the wall thickness is made too small in relation to the path length and to the area, there will be a danger of the metal cooling too rapidly (because the steel dies are good conductors of heat) and failing to spread over the whole area.

Suggested figures for wall thicknesses are given below:

Zinc pressure-die-casting alloys	0·6 to 2mm
Tin pressure-die-casting alloys	0·6 to 2mm
Lead pressure-die-casting alloys	1·0 to 2mm
Aluminium pressure-die-casting alloys	1·0 to 3mm
Magnesium pressure-die-casting alloys	1·0 to 3mm
Copper pressure-die-casting alloys	1·0 to 3mm

Wall thicknesses exceeding 4 to 5mm are not likely to be used.

Allowing for contraction.—Special precautions must be taken in view of the contraction occurring in pressure die castings. When parts are cast

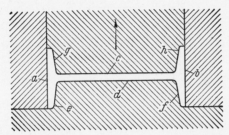

Fig. 102.—Allowing for contraction while in the dies

in mould boxes, the sand mould yields to the small changes in shape of the casting which take place during solidification, with the result that the casting does not stick in the mould. When a pressure die casting cools, however, the lack of give in the dies may cause the casting to stick fast if not suitably designed. During the cooling stage, the casting (fig. 102) will detach itself from the surfaces *a*, *b*, *c*, and *d* and shrink on to the surfaces *e*, *f*, *g*, and *h*. If the casting is held equally firmly in both halves of the die, there will be a risk of it breaking up during ejection. One method of preventing this is to use a core which retracts in the arrowed direction prior to the dies opening.

The casting should always shrink into the ejection half of the die. because it can be expelled from there. To enable shrinkage to take place in this way, the draft must be made larger towards the cover die.

Avoiding stresses.—A pressure die casting of the type shown in fig. 103a would be able to contract without restriction during cooling and would in fact detach its whole periphery from the die. If two holes are cast in the item however (as in example 103*b*), then the cores provided for this purpose will be bent if made too weak. This can be prevented by making the holes larger or by reducing the strain by using a large loose core (example 103*c*).

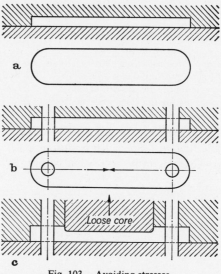

Fig. 103.—Avoiding stresses

Avoiding rambling shapes.—As in sand casting, items of rambling shape tend to distort in the arrowed direction (fig. 104). Shapes of this kind should therefore be avoided.

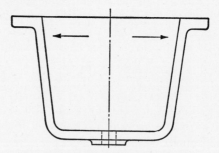

Fig. 104.—Distortion of rambling shapes

Avoiding distortion.—As in sand casting (fig. 49*a*), and for similar reasons, pressure die castings also distort if the wall thicknesses lack uniformity.

Providing for removal from die.—In pressure die casting, as in sand casting, it is necessary for surfaces perpendicular to the parting line and parallel to the direction of core withdrawal to be given a light draft (fig. 105). Those walls of the casting which detach themselves from the die on solidifying generally do not need to be given any draft. On the other

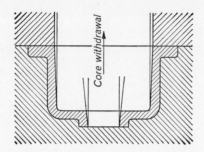

Fig. 105.—Allowing for removal from die

hand, the cores must always be given taper. The amount of draft given depends on:

1. The form of the casting.
2. The die casting alloy used.
3. The size of the area to be detached from the die.
4. The thickness of the walls.

The amount of draft applied is usually under 1 per cent, but when copper die-casting alloys are used it may go as high as 2 per cent.

Keeping the design straightforward.—One of the advantages of pressure die casting is that it permits the casting of extremely intricate shapes which

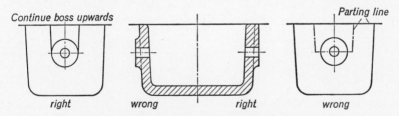

Fig. 106.—Designing for easy die parting

could scarcely be produced by any other casting method. In spite of this, however, the designer should generally aim to produce a design consisting of straightforward shapes which are easy to manufacture and which use straight parting lines as far as possible (fig. 106).

Avoiding cores.—The requirement that dies should be of simple design implies avoidance of cores (fig. 107). Hollow shapes always entail the use of cores. The designer will therefore aim to use ribbing as much as possible.

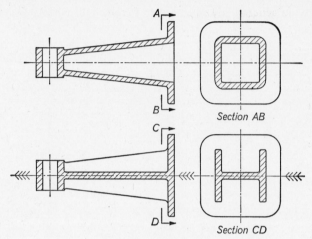

Fig. 107.—Avoiding cores

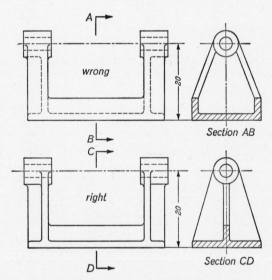

Fig. 108.—Correct form-design of a ribbed casting

But here, too, unsatisfactory disposition of the ribs may make coring necessary (fig. 108).

Avoiding closed cavities.—Closed cavities of the kind which can be made by sand casting are impracticable in pressure die casting (fig. 109). A casting of this kind is best split on the lines indicated in example *b*. This arrangement is also preferable for sand-casting purposes because it simplifies the problem of supporting the core.

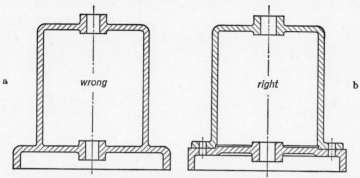

Fig. 109.—Avoiding closed cavities

Avoiding re-entrant features.—Re-entrant features in pressure die castings call for costly split cores. It is usually possible for the designer to avoid shapes of this kind when designing a component, and in this way he can reduce the cost of the dies considerably (fig. 110).

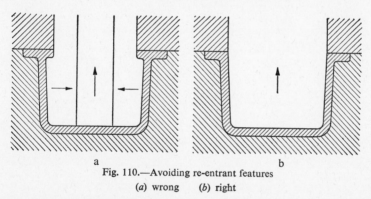

Fig. 110.—Avoiding re-entrant features
(*a*) wrong   (*b*) right

Allowing for machining.—Pressure die castings generally do not need any machining. Should machining be necessary, however, an allowance of 0·4 to 0·8 mm is sufficient.

Providing for easy flash removal.—A little of the metal injected may escape wherever parting lines occur. This causes flash to be formed. Intricate castings may exhibit quite a considerable amount of flash. The casting must therefore be so designed that the flash can be readily removed by trimming or stamping tools, since removal by hand is not economical. For this reason, the formation of flash in the interior of a casting is particularly undesirable. The simplest solution is to arrange for the flash

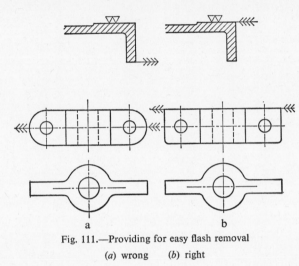

Fig. 111.—Providing for easy flash removal

(*a*) wrong   (*b*) right

to lie in a plane which will be machined, since it will then disappear automatically in the course of the machining operation (fig. 111).

Radiusing.—The provision of radii is governed in pressure die casting also by the position of the parting lines (fig. 112). Examples *c* and *d* call for multi-part dies and should therefore be avoided.

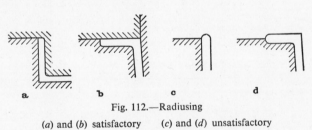

Fig. 112.—Radiusing

(*a*) and (*b*) satisfactory   (*c*) and (*d*) unsatisfactory

Standard of dimensional accuracy obtainable.—One of the advantages of pressure die castings is their high standard of dimensional accuracy. Inaccuracy is caused by the following factors:

Composite die construction using moving parts whereby play is introduced.
Difficulty in determining the contraction of the casting.
Thermal expansion of the dies.
Die wear.
Casting alloy used.

The following standards of dimensional accuracy are obtainable:

Pressure-die-casting alloy	Nominal size	
	15 to 100 mm	exceeding 100 mm
Zinc pressure-die-casting alloy Aluminium pressure-die-casting alloy Copper pressure-die-casting alloy	±0·15% ±0·20% ±0·25%	±0·10% ±0·15% —

The design of pressure-die-casting dies calls for at least as much knowledge and experience as does the design of the castings themselves. The designer therefore cannot be urged too strongly to discuss design problems with the tool engineer.

Finally, the principal factors affecting the design of pressure die castings are listed again below.

Rules

1. Aim at the simplest possible die design.
2. Avoid abrupt transitions.
3. Keep wall thicknesses as uniform as possible.
4. Avoid concentrations of metal.
5. Allow for contraction.
6. Allow draft where necessary.
7. Avoid partly closed shaped and closed cavities.
8. Make allowance for machined surfaces.
9. Provide for easy flash removal.
10. Take account of the standard of dimensional accuracy permitted.

(f) Form design of plastics mouldings

The synthetic plastic materials possess a range of outstanding properties. They can be readily processed without any need for cutting operations and turned into articles of extreme accuracy requiring hardly any finishing. It is true that the moulding method of manufacture only becomes economical if very large quantities are required (minimum of 1000). Runs of this order are much commoner in light engineering than in mechanical engineering in general. To this must be added the fact that (arising from the low strength values afforded) the mouldings concerned are comparatively small and of low strength, as would be naturally expected in light engineering. In this field a steadily increasing range of applications is opening up for synthetic plastics owing to the economy of manufacture they offer compared with metal die castings. In light engineering and in the construction of small machines, these materials have therefore assumed much greater importance than in mechanical engineering at large, although the parts in question must of course be capable of being processed mechanically without requiring cutting operations.

Attempts have recently been made to produce the entire bodywork for cars in plastic materials in order to secure more economical manufacture. The material used for this purpose is a ply-reinforced synthetic plastic consisting mainly of cotton fabric plus glass fibre and bonding agents. The manufacturing process is of course different from that employed in die casting. In England the production of bicycle frames in plastic materials started some years ago. It therefore seems very likely that in future plastic materials will also become increasingly important for highly stressed parts in general mechanical engineering.

To understand how to use plastics and design for them it is necessary to know all about their special properties. We will therefore examine the materials concerned more closely.

The various types of plastics have different properties determined by their chemical composition. They cannot be brought to a common denominator. Therefore, before he starts work on a design, the designer needs to make a close study of the materials available so that he can choose one suited to his requirements. For the purposes of mechanical engineering it is, of course, primarily the curing types which are of interest.

I. Physical properties

Specific gravity.—The synthetic plastics are the lightest constructional materials available to the machine designer. Their specific gravities average 1·4 to 1·8 and are therefore considerably lower than those of the aluminium alloys. Thus they can be used with advantage wherever a weight-saving method of construction is needed, as in automotive and aircraft engineering, always assuming that the necessary strength requirements are met.

Strength.—The question of strength properties is one of the utmost importance to the designer proposing to use a given type of plastic material. In addition to tensile strength and bending strength, it is also necessary with plastics to take into account the impact bending strength and notch toughness, as well as the elongation, if a more or less satisfactory picture of the practical usefulness of the material is to be obtained. The designer in general mechanical engineering is particularly interested in comparing the values with those of other materials whenever he is concerned with changing the material of which an existing article is made and going over to plastics instead. The properties of the high-grade plastics are approximately comparable with those of cast iron and the lower-strength light metals. It must be remembered of course that the elongation figures and elasticity of the plastics are much higher than those of cast iron.

Hardness.—The Brinell hardness of the plastics is generally very low.

Amounting on average to 40 to 60 kg/mm² (25-40 tons/in²) it is somewhat lower than the hardness of the aluminium alloys. For this reason, parts made of plastics cannot be used in situations where they will be exposed to damage through shock or impact.

Abrasion resistance.—All types of plastics offer very low abrasion resistance, and these materials are therefore unsuitable for duty involving heavy abrasion. Owing to their good sliding properties, the phenol and cresol resins are particularly good for bearing applications when used with a shredded cotton filler. Of course, an undesirable feature of these bearing materials is their poor heat-dissipating capacity which means in practice that the heat generated by friction can only be carried away via the lubricant circulation. The synthetic-resin bearing materials also absorb a little oil or water and swell as a result. The question of lubrication depends entirely on the operating conditions such as bearing pressure and speed of sliding. During the time that raw materials were controlled and rationed, the synthetic-resin bearing materials played an important part as substitutes for bearing metals. At the present day, however, it is probably true to say that the latter have in turn superseded the plastics to a large extent* though some plastics, e.g. nylon, hold their own.

Thermal properties.—Heat resistance is low for all plastics. There are some materials which must not be exposed to temperatures exceeding 60°C. There is also the fact that the linear thermal expansion is a multiple of that yielded by the heavy metals. This must be borne in mind by the designer whenever it is intended to reinforce plastics by moulding in metal parts. If the cross-sectional size of the insert parts is made too large, unduly high stressing may occur, and the plastic material may even be split off.

On the other hand, the designer can also make good use of the poor thermal conductivity of these materials by employing them, for example, as handles for heating or cooking appliances.

Electrical properties.—As will be well known, plastics are extensively used in electrical engineering owing to their poor electrical conductivity. This, however, is an aspect of their use which is outside the scope of the present book.

II. Chemical properties

Synthetic plastics are highly resistant to chemical attack. They are therefore used to great advantage in the chemical industry for pipelines, insulation and lagging, vats and casings. Owing to their high resistance to weathering no additional surface protection is needed. If immersed for a long time in water, however, they tend to absorb a little and swell. The

* Substitution of plastics for metals as bearing materials was perhaps more general in Germany than in England during the war years. Ed.

designer proposing to use plastic materials under these conditions must exercise caution.

III. Technological properties

The scope of the present book and the limited importance of these materials to the designer engaged in general mechanical engineering justify only a reference to the essential points of form design as it affects plastics, so that the designer can judge the practicability of applying them to his own work.

A point which should be noted at this stage is that successful form design of compression and injection mouldings demands cooperation between the design and production departments if for no other reason than that not all compression and injection moulding plants use the same methods and tools.

Proper use of the design possibilities offered by plastics demands knowledge of the properties of these materials and of the moulding techniques used. It is assumed that knowledge of the latter aspect of the subject has been gained in the course of technological studies. Moulding processes can be divided into three types, namely extrusion, compression, and injection.

The ordinary compression-moulding process has a number of disadvantages which are overcome by injection moulding. The latter imposes hardly any restrictions on the designs which can be produced. Owing to the uniformly through-heated condition of the particles of moulding material, the mouldings produced are homogeneous and free of stress. Furthermore, mouldings with non-uniform wall thicknesses are more effectively through-cured when moulded by the injection method. Dimenmensional accuracy also is better than that obtained with compression moulding. It is true, however, that these advantages are countered by certain drawbacks of an economic nature.

Owing to the high mouldability of practically all plastics, it is possible to mould small items in wall thicknesses down to 1 mm and even less. Typical values of wall thicknesses are as follows:

For small parts . . . .   1 to 2·5mm
For larger parts . . . .   3 to 5mm

As already indicated, the accuracy imparted to the mouldings produced is very high with injection moulding. The following standards of accuracy are obtainable over the distances indicated when the drawing specifies tolerances on dimensions:

Up to 6mm . . . .   ±0·1mm
From 80 to 120mm . . . .   ±0·4mm
From 400 to 500mm . . . . . .   ±1·5mm

Although the designer of plastics mouldings has a great deal of latitude, he should nevertheless observe certain rules. The most important of these are summarized below.

Rules

1. Aim to keep the shape of the moulding as straightforward as possible notwithstanding the great freedom in design.
2. To permit easy removal from the mould a draft of at least 1:100 is necessary on surfaces parallel with the direction of mould pressure.
3. Avoid inequalities in wall thickness as far as possible owing to the risk of stresses and cracking. Choose injection moulding if this requirement cannot be met.
4. Avoid thick-walled mouldings because curing is difficult or unsatisfactory.
5. Avoid sharp edges and corners owing to high notch sensitivity.
6. Avoid re-entrant features because these require expensive mould equipment owing to the removable members required.
7. Avoid hollow shapes. Where unavoidable, break down the design and stick the parts together.
8. Avoid concentrations of material. These will fail to cure properly and may cause cracking.
9. Use ribbing, not concentrations of material, to provide any necessary reinforcement.
10. Do not make critical working surfaces too narrow, for otherwise they may easily be damaged and rendered unserviceable owing to the low hardness of the material.
11. Moulded holes should have a maximum depth of two diameters to ensure that the moulding pin detaches easily. Deeper holes should be moulded offset.
12. Metal parts can easily be moulded in. Whether or not it is economical to do so, however, must be decided for each application on its merits. Points to watch in particular are:
    (a) that the plastic material surrounding the insert is ample in cross-section;
    (b) that recesses, notches, grooves, etc., are provided as a safeguard against twisting and pulling out;
    (c) that the inserts are properly located in the mould so that they do not become displaced under moulding pressure.

(g) Form design of welded fabrications

General.—Welding consists in uniting two separate metal parts of like composition to form a whole offering the maximum homogeneity. This result is achieved by a variety of processes the fundamentals of which must be assumed to be familiar to every designer.

Of all the joining processes—riveting, bolting, pressing together, etc.— welding is the ideal. To appreciate this, one need only compare the complicated stress pattern of a riveted or bolted fastening with that of a welded joint. In welding, the final product is always a unit which looks as if it has developed in one piece.

The big advantages of welding over other methods of joining metals were recognized very early. These advantages are to be found not only in great saving of material, low manufacturing costs, and dependability of the medium, but also in the fact that the designer is offered great latitude in planning and designing. It is therefore not surprising that one field of mechanical engineering after another has been opened up to the application of welding. Without welding techniques the lightweight methods of construction so vital to the automotive and aircraft industries would be unthinkable.

This does not mean, however, that in time welding will supersede all other joining techniques. It is a well-known fact that not all materials are satisfactorily weldable; exceptions, for example, are the high-grade alloy steels. Casting, too, has its own advantages to offer over welding. Apart from the lower manufacturing cost when long runs are involved, castings are also more readily machinable. A further point which should not be underestimated is the more attractive shape of castings.

Welding has about sixty years of development behind it. To begin with, designers had a certain distrust of this method of joining, particularly when welded joints were required to transmit considerable forces; there were no data available to allow accurate calculation. It is therefore easy to understand that it was at one time frequently said: welding is an art, but riveting is a science. Over the past two decades, however, welding processes have been developed to such a standard that structures fabricated by their aid are now also amenable to calculation and can stand up to exacting strength requirements.

Welding processes.—The processes used in welding can be divided into two groups, as follows:

1. Pressure welding in which the parts are joined in a plastic state by the application of pressure.
2. Fusion welding in which the individual parts are united in the molten condition without the application of pressure.

These two groups of processes can be further subdivided as follows:

1. *Pressure welding:*	2. *Fusion welding:*
(a) Hammer welding	(a) Gas welding
(b) Electric-resistance welding	(b) Arc welding
(c) Thermit pressure welding	(c) Thermit fusion welding

Of the above methods, the ones of special interest to the designer are the electric-resistance methods known as *resistance-butt welding* and *flash welding* which are used for fabricating shaft components often differing in strength (fig. 113) and for welding fork ends, etc., on to rods (fig. 114).

The great advantage of resistance-butt welding is that the strength of the welded joint is virtually the same as that of the parent material (90–100 per cent).

For sheet fabrication purposes, where only tacking is required, *spot welding* is a profitable method for use on steel under 10 mm thick and on aluminium under 3mm thick. When a continuous uninterrupted weld is required, however, the process used is *seam welding* in which the electrodes

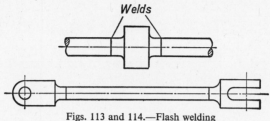

Figs. 113 and 114.—Flash welding

are in the form of rollers. Seam welds may be either lapped (fig. 115*a*) or made flush by the methods shown in *b*, *c*, *d*, and *e*, but only for sheet thicknesses up to 2·5mm. (Steel radiators, for example.)

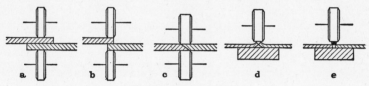

Fig. 115.—Seam welding

Gas or arc welding.—The two principal welding methods and the ones most widely used are *gas welding* and *arc welding*; the question of which method to employ is not always easy to answer. It can be said with certainty, however, that during the past few decades arc welding has become more and more widely established. This factor is linked with the big advantages it offers, since the temperature of the electric arc is much higher than that of a gas flame so that, in contrast to gas welding, the joint zones melt practically instantaneously. With gas welding, the long pre-heating period also raises the metal adjacent to the joint to a high temperature and thereby exerts an unfavourable effect on the crystalline structure. This in turn may lead to considerable stresses being set up. Gas welding is unsuitable for relatively large cross-sections owing to the trouble and time involved in pre-heating. Plates above 20mm thick are therefore best welded electrically. One result of the high temperature attained in electric welding is the fact that the welding operation is accomplished more quickly and therefore more economically.

As to which of these two welding processes is the more economical, the answer is that for small firms working at less than full capacity, gas welding is more economical than the electric processes, so long as the work to be welded is less than 6mm thick. In addition, gas welding is the method which can be used for welding all metals capable of being welded. It is still used to great advantage for welding all the non-ferrous metals.

The atomic-hydrogen welding process has found wide application. It resembles gas welding and is suitable for sheet and plate from 1 to 80 mm thick. Preparation of the edges to be welded is the same as for gas welding. It is not very suitable for welding cast iron but is quite satisfactory for all non-ferrous metals. When steel is welded by this method, the welds produced offer great strength and elongation.

For the series production of welded parts, machines have been developed on the " Unionmelt " principle in which the welding rod is fed automatically and moved forward mechanically at a constant rate. In this way the welding process is carried out more rapidly and the weld face is left smooth. This method is particularly suitable for heavy work (plate thicknesses ranging from 5 to 60 mm). With this method there is no need to prepare the plate edges by bevelling.

Weld forms.—One of the big advantages of welding is that it allows metal parts to be fastened together very simply without the aid of straps, laps, or angles. The weld forms used can be classified as butt welds, fillet welds, and miscellaneous welds.

1. *Butt welds.*—Plate up to about 3 or 4 mm thick can be welded without any edge preparation; all that is necessary is to leave a small gap between the parts at the welding joint. When welding thicker plate it is necessary to provide an enlarged groove by bevelling the edges of the plates at an angle of 60 to 80° (fig. 116*b*, *c*, *d*, *e*, and *f*).

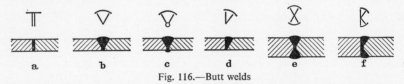

Fig. 116.—Butt welds

2. *Fillet welds.*—Fillet welds are less satisfactorily stressed and therefore weaker. Three types are distinguished (fig. 117*a*, concave fillet weld; *b*, mitre fillet weld; and *c*, convex fillet weld).

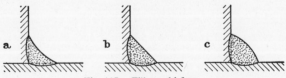

Fig. 117.—Fillet weld forms

In spite of the larger cross-section of the convex weld form, the latter is little used because it has the most pronounced notch effect. The most satisfactory form in this respect is the concave fillet weld. In fillet welding it is often unnecessary to make any special preparation at the joint (fig. 118*a* and *b*). By means of edge preparation, however, it is possible to

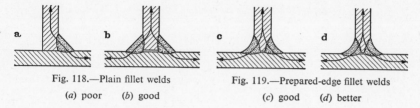

Fig. 118.—Plain fillet welds

(*a*) poor   (*b*) good

Fig. 119.—Prepared-edge fillet welds

(*c*) good   (*d*) better

improve the stress pattern and, consequently, the strength of fillet welds (fig. 119*c* and *d*).

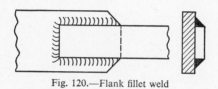

Fig. 120.—Flank fillet weld

Fig. 120 shows a flank fillet weld which is equivalent to a butt weld under conditions of static loading. Considered three-dimensionally, however, the stress pattern is very much less satisfactory than that yielded by a butt weld. At the ends of the welds there occur notch stresses which can be lessened to some extent by machining off the end craters. Butt welds are to be preferred if alternating stresses are to be withstood.

A fillet weld stressed in torsion is referred to as a ring weld (fig. 121).

3. *Miscellaneous weld forms.*—Sheet less than 1·5 mm thick cannot be satisfactorily butt welded. The method used for thin sheet is to turn up the

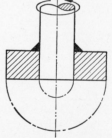

Fig. 121.—Ring weld

edges (fig. 122*a*) and make a flush joint by gas welding after applying a suitable flux (fig. 122*b*).

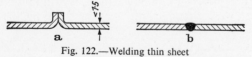

Fig. 122.—Welding thin sheet

Slot welds and plug welds (fig. 123*a* and *b*) are of minor importance and are best avoided.

Welding offers the designer great latitude. Nevertheless, he must understand and allow for the unique physical phenomena occurring in welding, so that he will be able to make the right choice from among the

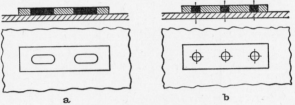

a   b

Fig. 123.—Slot welds and plug welds

many different types of joint available. Brief mention will therefore be made of the most important of these factors.

Fig. 124

It is not always possible to avoid making joints between plates which differ in thickness. Provided that the difference is not too large the type of joint shown in fig. 124 can be used.

When bigger differences in plate thickness are involved (fig. 125*a*, *b*, and *c*) welding can only be carried out by the electric-arc method owing to the more concentrated heat input obtained thereby. Because of the

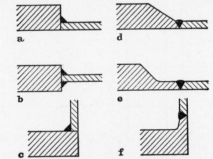

Fig. 125.—Welds used when plate thicknesses are unequal

notch effect, however, this type of joint also is only suitable for static loading. Alternating loading calls for gradual changes in cross-section. This requirement can be met by the methods shown in fig. 125*d*, *e*, and *f*.

For gas-welding purposes, the methods shown in fig. 125*d*, *e*, and *f* are essential.

Stresses.—Very undesirable effects on the work can result from the contraction of a weld during and after welding. Before dealing with this in more detail we will examine the types of stress with which the designer is most likely to be concerned.

The stress concerned can be considered under three headings: namely initial stress, expansion stress, and contraction stress.

Initial stress.—Rolled sections and plates are affected by initial stress arising through non-uniform cooling after the rolling process. Castings also often have initial stress for the same reason. The action of working such materials may disturb the equilibrium between the internal forces. This leads to objectionable distortion in the component concerned. (Examples are to be found in cutting keyways in shafts and in planing castings.)

Expansion stress.—A very simple way of showing the occurrence and the consequences of expansion stress is by means of a frame (fig. 126) having a crack at the point *a*. If gas welding is used, the metal on either side of *a* will be intensely heated so that expansion will take place in the arrowed directions. If the metal is of a brittle nature, such as grey iron for example, cracks may be formed at *b* and *c*. With arc welding this risk might well be lessened owing to the reduced local heating. The remedy is to apply simultaneous heating at *d* or to carry out the welding slowly with suitable pauses allowed.

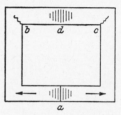

Fig. 126.—Setting up of expansion stresses

Contraction stress.—For a correct appreciation of the nature of contraction stress it is necessary to note the following point. If a shaft constrained between two walls is heated (fig. 127*a*) it will expand in all

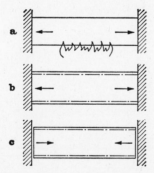

Fig. 127.—Setting up of contraction stresses

directions. If this expansion is obstructed in the longitudinal direction the shaft will be upset, provided that it is thick enough not to buckle. After the shaft has cooled it is found that it has become shorter but thicker (fig. 127*c*).

Similar conditions apply in welding (fig. 128). The welding of a crack in a plate, for example, will heat up the hatched area and cause it to expand. Since it is prevented from doing this, however, it will become

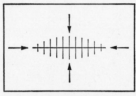

Fig. 128.—Contraction stresses parallel and at right angles to
line of weld

thicker. Thin sheet of course will bulge out. After it has cooled, the metal previously heated will contract so that tensile stress (contraction stress) will be set up in the arrowed directions.

In the weld itself, three-dimensional contractions are set up during cooling and these are effective in all directions, namely parallel with the

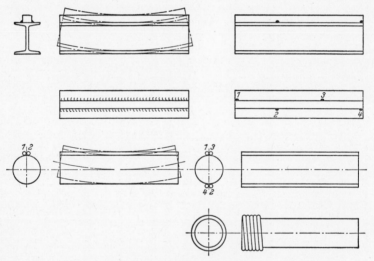

Fig. 129.—Consequences of longitudinal contraction. The remedy
is shown on the right

weld, at right angles to the weld, and parallel with the thickness of the plate.

The consequences of longitudinal contraction are warping, curling, and twisting (fig. 129).

In the example shown in fig. 129 a further consequence of longitudinal contraction can be seen. During welding, the heat travels along the rail in advance and lengthens it, so that contraction is further increased by a considerable amount on cooling.

Contraction at right angles to the weld causes the two surfaces joined to be pulled together. If this is prevented, severe tensile stresses are set up.

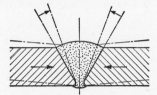

Fig. 130.—Consequences of crosswise contraction (angular contraction)

With a single-V butt weld the top will contract more than the bottom (fig. 130) owing to the greater thickness of the weld at the top. Angular contraction therefore occurs.

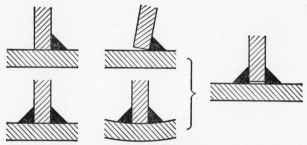

Fig. 131.—Consequences of crosswise contraction. The remedy
is shown on the right

Some of the characteristic effects of crosswise contraction are illustrated in the examples given (fig. 131).

When plates which are free to move are welded together (fig. 132a) they tend to approach each other as shown in b owing to crosswise contraction. The remedy for this is to prepare as shown at c.

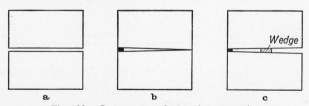

Fig. 132.—Consequence of crosswise contraction

Measures to combat contraction stress.—There is a whole range of techniques which can be used to minimize contraction stress. Although these measures are entirely concerned with the production side, the designer must nevertheless know about them and may even refer to them in works drawings.

The most important of these measures are the following:

1. The ideal way of eliminating stress is by means of subsequent annealing. When the work is large, however, this may be difficult to carry out.
2. The work is heated before or during welding to bring about a redistribution of stress between the weld and the work.
3. Hammering of the weld itself (to spread it) and of the adjoining zones during welding has the effect of reducing contraction.
4. Hammering after welding on and alongside the weld also has the effect of reducing contraction.
5. By welding on the step-by-step principle the tendency for the work to twist round at right angles to the weld is avoided (fig. 133).

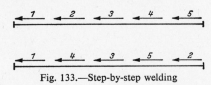

Fig. 133.—Step-by-step welding

Welding sequence.—The designer cannot begin to design successfully for welding until he appreciates the harmful effects of contraction stress and minimizes them by appropriate design methods. Of outstanding importance in this respect is the welding sequence. Unintelligent assembly may result in the setting up of severe contraction stress so that the work

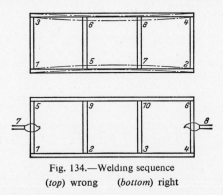

Fig. 134.—Welding sequence
(*top*) wrong   (*bottom*) right

distorts. For this reason, the designer must consider the most favourable welding sequence to adopt and inform the workshop accordingly by means of a suitable plan. A simple example (fig. 134) is sufficient to illustrate the effect on the work of an incorrect welding sequence.

Force pattern in butt welds.—For many applications welding has superseded riveting. The reason for this is easy enough to understand when one stops to consider just how ideal are the joining methods offered by welding compared with riveting. An example will illustrate this. The joining together of two flat steel bars subjected to a tensile load can be done in three ways by riveting (fig. 135).

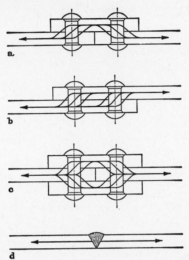

Fig. 135.—Force pattern obtained in riveted and welded joints

The stress distribution indicated for each of the three methods shows quite a complicated pattern with bending stresses added to the tensile stresses. Under the influence of the pull applied, the riveted joints *a* and *b* try to deform in such a way that the lines of force take on the minimum of curvature.

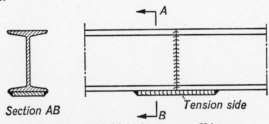

Fig. 136.—Welded joint between two H-beams

With welding, the bars can be butted together to give straight-line forces which give no tendency to deformation and which also ensure optimum utilization of the metal in the joint (fig. 135*d*).

In the making of welded joints, therefore, the aim will be to avoid all laps and straps. The only exception to this is when the strength of the weld in itself is not sufficient (fig. 136).

With fillet welds the force pattern is not so favourable as with butt welds. Experiments have also shown that lower strength figures must be assumed for the former. It should therefore always be the designer's aim to manage with butt welds whenever high strength is needed. The welded boilers used in steam locomotive construction are an example of this.

Corner joints and T-joints call for angle pieces at the corners when riveted (fig. 137*a* and *b*). With these configurations also the force pattern

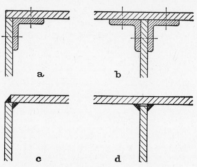

Fig. 137.—Corner joints and T-joints

is very complicated. With welded structures the angle pieces used for stiffening purposes can be dispensed with so that a much simpler force pattern is obtained (fig. 137*c* and *d*).

The strength obtained in the weld itself is always lower than that of the parent metal. This explains why the designer should avoid placing welds at vulnerable points (fig. 138).

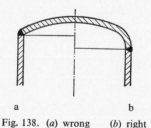

Fig. 138. (*a*) wrong   (*b*) right

Joints used in welding.—Before he approaches the task of designing a structure that is fit for welding, the designer should make sure that he knows the different methods of joining available, since these not only have a bearing on the fabrication and strength of the structure, but also influence its entire shape.

Corner joints.—The types of corner joint shown in fig. 139a and b are used only with gas welding processes, whilst c and d require fitting and thereby differ from e and f. Types c, e, g, h, i, and k are not strong when subjected to bending. Types i and k are forms of edge weld used for sheet under 3mm thick.

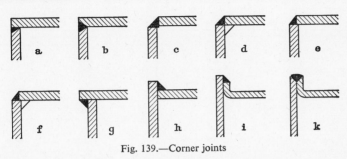

Fig. 139.—Corner joints

Single-T and H-joints.—The type shown in fig. 140a is a corner joint made with a fillet weld. A better version is b because the weld form acts like a butt weld. Types c and d are good, but e and f are the best from the

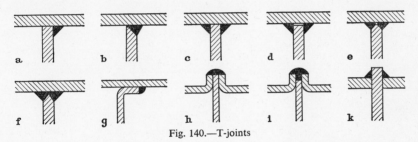

Fig. 140.—T-joints

strength point of view. Type g is used only when joining a thin sheet to a thicker one. Types h and i are edge welds with an intermediate thin sheet.

Flanges.—The type shown in fig. 141a is certainly the best, but is

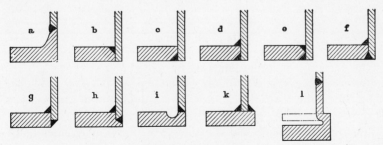

Fig. 141.—Flange joints

expensive; types *b* and *c* are suitable only for light loading. Examples *d*, *e*, *f*, *g*, and *h* offer good strength. Type *i* is suitable for gas welding. Types *k* and *l* are used for connections to tubeplates.

Reinforcement.—Reinforcement is made by the methods shown in fig. 142*d* and *e*; *a*, *b*, and *c* are methods of fabricating eyes and bosses by welding.

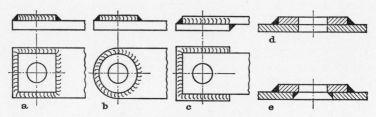

Fig. 142.—Methods of reinforcing

Corner joints for rolled products.—Sheet, plate, and steel shapes are trimmed by guillotining and sawing either manually or by machine, and also by flame-cutting. The simplest shapes for cutting are shown in fig.

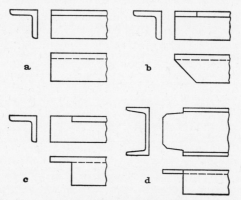

Fig. 143.—Machine-trimmed ends

143*a* and *b* for butt and mitre joints respectively. Alternatively, the webs can also be cut away as in *c* and *d*.

For examples of corner joints see fig. 144.

Corners of tanks and containers.—The types of corner joint shown in fig. 139*c* to *k* are practicable from the welding and strength points of view.

Sharp edges on tanks and containers are always more notch-sensitive than rounded edges. In addition, it is the edges which are most vulner-

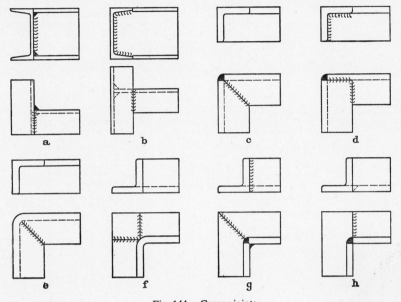

Fig. 144.—Corner joints

able. Type 145*c* is of course dearer than *a* and *b*, but is better, quite apart from the more pleasing shape.

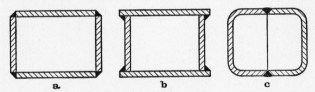

Fig. 145.—Box sections

Ribs.—Tanks and containers with flat walls require stiffening if subject to pressure. The necessary reinforcement can be provided on the lines shown in fig. 146.

Fig. 146.—Types of rib

Bolted connections.—The designer must give special attention to points where forces are set up by bolted fastenings. The necessary reinforcement can be provided in the manner shown in fig. 147.

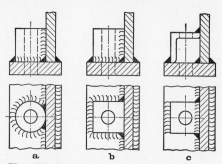

Fig. 147.—Reinforcement for bolted fastenings

Hub joints.—Drums, wheels, etc., fabricated by welding will, of course, also have the hubs attached to the discs or spokes by welding. Some examples of the methods used are given in fig. 148.

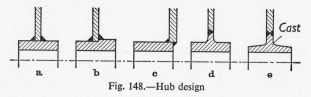

Fig. 148.—Hub design

Type *a* is the simplest method of attachment. The shoulder shown in type *b* makes for easier fabrication. Types *d* and *e* are favoured when cast-steel hubs are used.

Rod ends.—When only small quantities are required, the designer is very often faced with the decision whether to forge or weld the item in

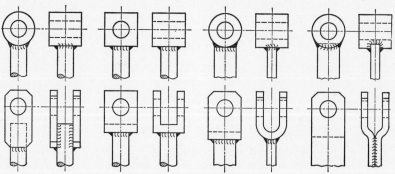

Fig. 149.—Welded rod ends

question. The examples show methods of fabricating rod ends by welding (fig. 149).

Pipe joints.—Welded pipe joints are a subject in themselves. Gas welding offers advantages for this type of work. The examples in fig. 150 show

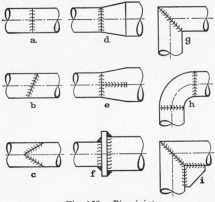

Fig. 150.—Pipe joints

longitudinal joints and corner joints. The weld form used depends on the thickness of the pipe wall.

Fig. 151 shows methods of making welded branch joints in pipes. Types *b* and *c* provide the greatest strength and the most satisfactory flow conditions. A useful method when several pipes meet at the same point is to run them into a sphere.

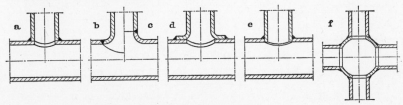

Fig. 151.—Pipe branches

The design of tubular structures, strutting, and bracing often results in involved sections (fig. 152*c* and *d*) which call for appropriate trimming and preparation.

Weldability of steels.—Not all steels are weldable. For unalloyed steels a carbon content of 0·3 per cent usually represents the limit for reliable welding.

As far as high-alloy steels are concerned, information on weldability can be obtained from the relevant data sheets. With the exception of the

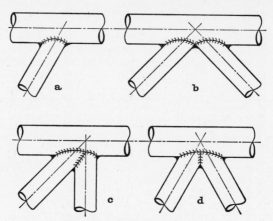

Fig. 152.—Diagonal connections

martensitic and ferritic grades, these steels are usually readily weldable with suitable electrodes.

For the fabrication of welded structures there are available to the designer:

1. Plates to B.S. 1449: 1956
2. Rolled sections to B.S. 4: 1962
3. Various lightweight sections including non-circular tubes.

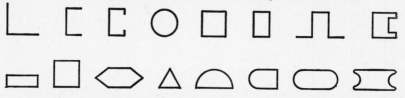

Design of welded structures.—Welded structures can be designed to be:
1. Fabricated entirely from plate (fig. 153), or

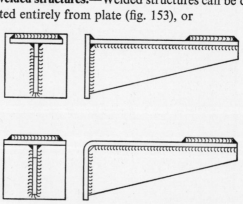

Fig. 153.—Brackets fabricated from plate

2. Built up from plate and rolled sections (fig. 154).

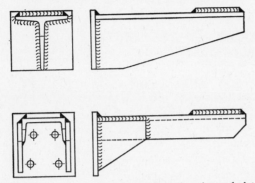

Fig. 154.—Brackets fabricated from rolled section and plate

For large-sized machine components such as columns, frames, etc., the following methods of construction have been developed over the years:

1. Plate construction
2. Built-up plate construction
3. Hollow construction
4. Cellular construction

Plate construction.—This method utilizes the extremely high bending strength of plates standing on edge, and is the simplest method of all.

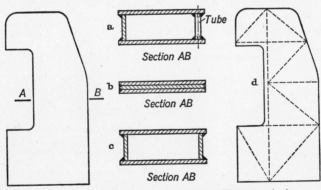

Fig. 155.—Machine frame constructed by different methods

Cross-stiffening is provided usually by means of round bars or tubes. It is not a suitable method for structures required to resist vibration (fig. 155a).

Built-up plate construction.—This is closely related to the plate construction method. It is suitable only for resisting bending loads. The method consists in welding together the outer and inner edges of several plates. The welds are intended to carry any load differences between the plates (fig. 155*b*).

Hollow construction.—This method gives rise to forms which are closest to those used in the design of castings. It has the advantage of offering extreme stiffness to bending and torsional loads (fig. 155*c*).

Cellular construction.—This method is used whenever a certain amount of rigidity is called for in addition to extreme resistance to bending and torsional loads (fig. 155*d*).

Welded structures of this kind can even be built as rigid as castings, as demonstrated by many experiments.

The main factors to be borne in mind in designing welded structures are summarized below.

Rules

1. Do not attempt to copy blindly from cast, riveted, and forged designs.
2. Provide for a straight-line force pattern as far as possible.
3. Avoid laps, straps, and stiffening angles.
4. Use butt welds wherever possible.
5. Limit the number of welds used.
6. Make sure that ends to be welded together are of equal thickness.
7. Avoid placing welds in vulnerable cross-sections.
8. Facilitate assembly by means of registers, shoulders, etc.
9. Avoid the use of welding fixtures as far as possible.
10. Provide for easy access to welds.
11. Allow for the effects of thermal stresses.
12. If alternating stresses are involved, avoid running a weld at right angles to the direction of maximum principal stress owing to the low fatigue resistance offered by welds.
13. Distribute heavy loading over long welds in the longitudinal direction.
14. Avoid subjecting welds to bending loads.
15. Do not site a weld at the point of maximum deformation.
16. Be careful with the use of ribs. Incorrectly designed and dimensioned ribs will lead to undesirable intensification of notch effects.
17. Avoid large flat walls which tend to bulge and flex. Use swages; ribs are not always advantageous.
18. Consider the order in which parts shall be welded together and prepare a plan accordingly for the workshop.
19. Make sure that works drawings contain the necessary information, such as details of weld quality, weld form, weld length, etc.

Exercise problems

Problem 19

Try to design the plain cast-iron bearing bracket (fig. 156) in various ways suitable for welding.

Problem 20

It is required to attach a pin to a channel section as shown in fig. 157 by means of a cast flange and by welding so that the pin is fixed but detachable.

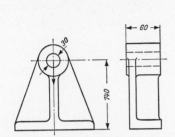

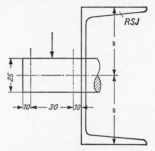

Fig. 156.—Small bearing bracket (cast iron)   Fig. 157.—Attachment of a pin

Problem 21

It is required to replace the familiar type of cast-iron hanger (fig. 158) by a welded fabrication. Keep to the dimensions specified.

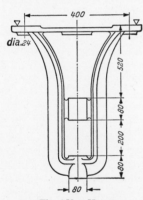

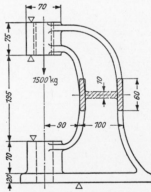

Fig. 158.—Hanger   Fig. 159.—Gap press frame

Problem 22

In what different ways could the gap press frame in cast iron as shown in fig. 159 be designed as a welded fabrication?

Problem 23

The bedplate shown in fig. 288 is designed to be torsionally stiff. How should it be designed for fabrication by welding?

(h) Form design of forgings

The principle of forging.—There are some metals which at elevated temperature have the property of assuming a plastic condition in which they can be shaped, i.e. forged, by hammering or pressure without undergoing any appreciable loss of mechanical properties.

Processes.—The forging process can be divided into two types, namely:

1. Hammer forging, in which the desired shape is imparted by hand, or mechanically, more or less by eye, and
2. Drop forging in which dies made of cast iron, steel, or cast steel are used in order to give maximum accuracy of shape to the forgings produced.

Hammer forging.—It is not within the scope of a book of this kind to discuss the technology of forging; instead, it must be assumed that the student designer already possesses the necessary knowledge. He needs to know all about the tools, machines, and auxiliary appliances used by the blacksmith, and must be familiar with all the methods of working employed, such as drawing, widening, upsetting, piercing, slotting, cutting, stepping, bending, and twisting. The designer's task is to use this knowledge, so as to impart to the forging the most satisfactory shape possible, so that it will be best fitted to the special nature of the forging process.

Advantages of forging.—The advantages offered by forging are considerable and are listed again below.

1. Drawing and widening compact the structure and reduce the grain size so that the strength and toughness of the metal are increased. Stresses cannot arise while still in the unloaded condition.
2. The action of drawing causes the fibres to run parallel with the boundary of the work, whereas the fibres are cut when the item is machined from the solid (fig. 160*a*). A forging is therefore stronger than an item produced by machining.

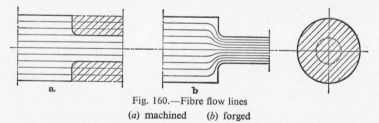

Fig. 160.—Fibre flow lines

(*a*) machined   (*b*) forged

3. From the examples given it will be seen that forging not only imparts greater strength but also secures substantial savings in material compared with manufacture by machining methods.
4. When small quantities are involved, hand forging saves time and enables the work to be produced more cheaply than by casting.

Selection of material.—The problem of choosing suitable material has already been dealt with, starting on page 54. Naturally, the critical requirements to be met are those arising from the function required and from the conditions in service. Whenever doubt arises, however, as indeed in all problems concerned with the choice of material, it is advisable for the designer to work in close cooperation with the forging specialist.

Principles of form design.—The correct form design of forgings requires of the designer that he should be fully acquainted with the nature of forging techniques. Only the more important factors will be discussed here.

Forging or welding.—Before starting work on the design of a forging, the designer should always ask himself if it is absolutely necessary for the

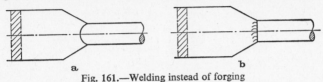

Fig. 161.—Welding instead of forging

workpiece in question to be produced by forging. Very often it will be found that the same result can be obtained by other manufacturing methods. For example, it is not always necessary to use forging in order to make a collar on a shaft. Often it will be found that a welded-on or shrunk-on ring is adequate for the purpose.

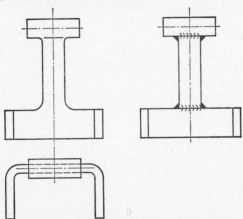

Fig. 162.—Welding is often cheaper than forging

Transitions from rectangular to circular cross-sections can of course also be made by welding (fig. 161*a*). Fabrication by welding is indeed more straightforward and cheaper.

The designer should always consider whether it will be possible to cut manufacturing costs by means of welding. He will often find that this is so (fig. 162), unless some specific shape, or the need for high strength, demands manufacture by forging.

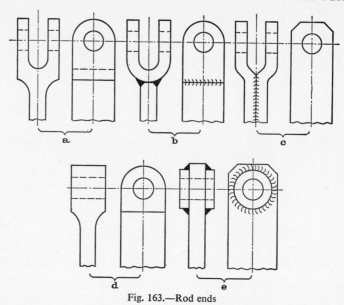

Fig. 163.—Rod ends

Fig. 163*a* and *d* show forged rod ends which could be replaced by welded versions.

Fig. 164 is an example of flash welding which offers the great advantage that the strength of the butt weld is nearly equal to that of the adjoining metal.

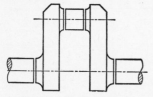

Fig. 164.—Flash welding

Choosing the simple shape.—By correct choice of shape, the designer can reduce the amount of deforming to be carried out during welding, and can thereby lower the cost of manufacture (fig. 165*b*).

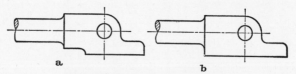

Fig. 165.—Choosing the simple shape

(*a*) unsatisfactory   (*b*) satisfactory

Allowing for the special character of the forging process.—The student designer cannot be urged often enough to make allowance for the special character of the manufacturing methods he specifies. It would be a mis-

take, for example, to attempt to make forgings as replicas of malleable
iron castings or steel castings (fig. 166).

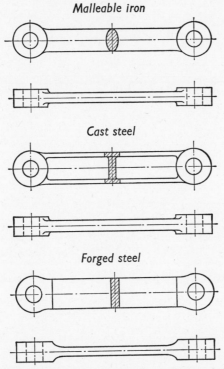

Fig. 166.—Influence of manufacturing method on design

Avoidance of upsetting.—Drawing out is a simpler method of making
a collar than upsetting, particularly if the collar in question is situated in
the middle of a rod. Collars at the ends of rods can be made by upsetting.

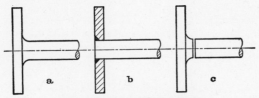

Fig. 167.—Avoidance of upsetting

Large flanges on the ends of rods (fig. 167*a*) should not be made by upsetting; instead, they are best attached by riveting or welding (fig. 167*b*
and *c*).

Avoiding steep tapers.—The first stage in tapering a bar is to set it down in a series of steps by hammering (fig. 168a). A flatter is then used

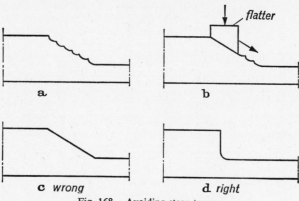

Fig. 168.—Avoiding steep tapers

to smooth the steps (*b* and *c*). It is easy to see that if the slope is too steep the flatter will skid off when struck. For this reason it is best to avoid

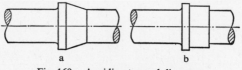

Fig. 169.—Avoiding tapered diameters

(*a*) wrong   (*b*) right

steep tapers and to adopt instead gentle transitions or sharp shoulders (type *d*).

For the same reason it is also preferable to avoid tapered diameters and to replace them by parallel transition diameters (fig. 169).

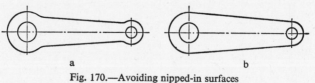

Fig. 170.—Avoiding nipped-in surfaces

(*a*) wrong   (*b*) right

Avoidance of nipped-in surfaces.—Nipped-in surfaces involve more work than plain shapes. The designer with an eye to economy will therefore avoid them (fig. 170).

Abrupt transitions.—Abrupt transitions are more difficult to forge and for strength reasons alone should be avoided (fig. 171).

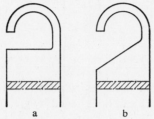

Fig. 171.—Avoiding abrupt transitions
(*a*) wrong   (*b*) right

Round bosses.—The forging of a bar to a more or less circular cross-section calls for the use of a swage. The same applies to the forging of round bosses. The designer should therefore avoid asking too much of the blacksmith, and should prepare his design in such a way that the smith will not have to use auxiliary appliances. Therefore, as far as forgings are

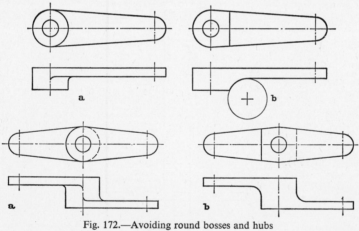

Fig. 172.—Avoiding round bosses and hubs
(*a*) wrong   (*b*) right

concerned, the designer should avoid round bosses (fig. 172*a*) and instead should set them down straight, so that ordinary tools can be used to produce them (fig. 172*b*). Type *a* would be more difficult to machine after forging than type *b*. Similar considerations also apply to the form design of the boss for the link (fig. 172).

The shapes shown in fig. 172 are easily made as castings or drop forgings.

6   (H 642)

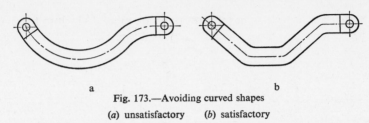

a   b

Fig. 173.—Avoiding curved shapes

(*a*) unsatisfactory   (*b*) satisfactory

Curved shapes.—Curved shapes entail a lot of troublesome matching to a template. This work can be lessened by giving the forging a suitably simplified shape (fig. 173*b*).

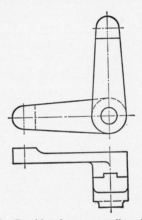

Fig. 174.—Breaking down a complicated forging

Difficult forgings.—By breaking the item down into two or more parts it is often possible to simplify the forging work and to manage with smaller blanks (fig. 174).

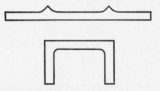

Fig. 175.—Avoiding sharp corners

Sharp corners.—Sharp corners should always be avoided on the curved parts of a forging unless obligatory for design reasons. Corners of this kind can of course be produced, but they are expensive because they require extra forging work (fig. 175).

Bending radii.—When a workpiece is bent, the inner fibres are compressed and the outer ones elongated. If the radius of curvature is made too small, it may happen that folds will be caused on the inner side and cracks on the outer side, particularly if the work is bent at right angles to the direction of rolling. Therefore, the rule applying to all bending work

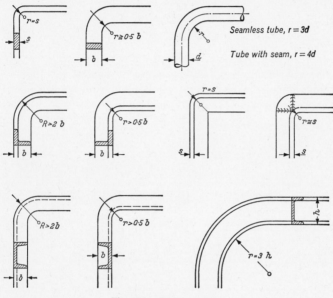

Fig. 176.—Bending work

is that bending should only be carried out parallel to the direction of rolling. Bending can be performed with the work cold or hot. Smaller radii of curvature are obtained when the work is hot.

The smallest bending radii which can be allowed are determined by the type of metal and its thickness. The following figures can be taken as a very rough guide:

Soft metals, minimum $r = 0.5s$ (s = thickness)
Hard metals, minimum $r = s$
Spring-hard metals, minimum $r = 3s$

The minimum allowable bending radii for rolled sections in common use are given above (fig. 176).

Bends giving too tight a transition are difficult to make (fig. 177*a*). Care should therefore be taken to provide a perpendicular run-in as at *b*.

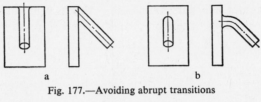

a   b

Fig. 177.—Avoiding abrupt transitions

(*a*) wrong   (*b*) right

Forging and welding.—A combination of forging and welding can often simplify manufacture. The bell crank illustrated would be difficult to make if forged throughout (fig. 178*a*). By welding the fork on to the previously

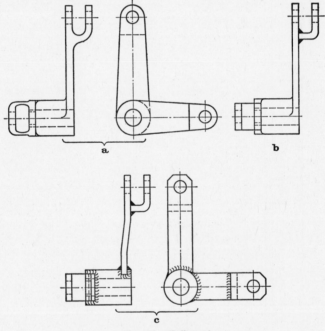

Fig. 178.—Bell crank

forged arm, however, the cost of manufacture can be cut (*b*). With a design like this, a further point to consider is whether it would not be more economical to treat the entire crank as a welding proposition (*c*).

Machining allowances.—The designer must also provide the necessary machining allowances in his design. These allowances are as follows:

Small items	3 mm
Medium-sized items	5–10 mm
Large items	15–30 mm

Tolerances.—The designer must prepare for the blacksmith's shop a special drawing in which the finished dimensions, the specified dimensions and the machining allowances are entered (fig. 179), including their

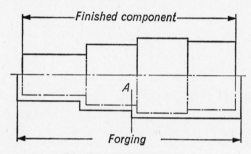

Fig. 179.—Forging tolerances

respective tolerances, as marked off from a datum edge A. The question of the degree of accuracy to be specified should be considered with great care. An unduly tight specification in this connection will add to the cost of the forging.

The factors to be considered in the form design of a hand forging are summed up below.

Rules

1. Make a careful check to see whether forging work can be replaced by welding.
2. Allow for the special characteristics of forging.
3. Choose simple shapes.
4. Avoid abrupt transitions.
5. Avoid upsetting as far as possible.
6. Avoid steep tapers and tapered diameters.
7. Avoid nipped-in surfaces and cavities.
8. Avoid round bosses.
9. Avoid curved shapes.
10. Break down difficult forgings.
11. Avoid sharp corners.
12. Do not make bending radii too small.
13. Specify machining allowances.
14. Specify forging tolerances.

Exercise problems

Problem 24

The crank (fig. 180) and the forked lever (fig. 181) have been taken from a very old book dating from the last century. It is stated in the book that both of these machine components are of forged steel machined all over.

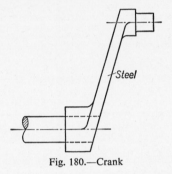

Fig. 180.—Crank

To what extent is the shape of the two items open to objection from the forging point of view, and what suggestions can be made to improve the economic aspect of the design?

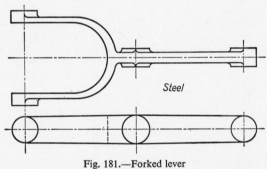

Fig. 181.—Forked lever

Problem 25

The steam engines built at one time with what were known as back-acting connecting rods used crossheads of the type shown in fig. 182. These were forged from steel and machined all over.

How should the crosshead be designed as a forging if it is proposed to machine only the bearings?

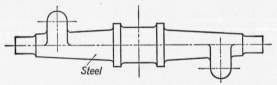

Fig. 182.—Crosshead

Problem 26

The frame-type valve-operating rod shown in fig. 183 was originally made by forging.

Examine the various stages of the work and illustrate them by freehand sketches.

How would it be possible to make this machine component more easily and more cheaply?

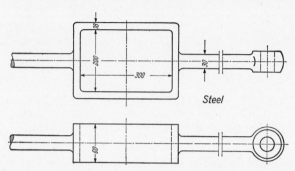

Fig. 183.—Frame-type valve-operating rod

Drop forging.—In contrast with hammer forging, in which the work is freely shaped, the principle of drop forging is to use impact or pressure to work the metal to its finished shape in die impressions.

Owing to the high cost of the dies this method of manufacture is only practicable when large quantities are required. The economic practicability of drop forging can only be assessed by making a detailed calculation (see page 166).

Compared with hammer forging, drop forging offers big advantages. The forgings can be made to much closer tolerances, so that they need little or no finishing apart from flash removal. The forgings produced are therefore interchangeable. Provided that the quantities are sufficient, the cost of manufacture is considerably lower than for a hand-forged item.

As far as the designing of work for drop forging is concerned, there are certain factors to be noted by the designer. The technique of drop forging, however, depends on so many kinds of experience, such as the conditions required to obtain proper flow of the metal, the most satisfactory form to give to the dies, the question whether to use drop forging or press forging, etc., that the acquiring of this knowledge alone can be a lifetime's task for the operator.

The designer faced with a difficult problem should therefore discuss it with the forging expert before attempting to evolve a design suitable for drop forging.

Influence of forging method.—Even the type of forging method used influences the design. Forgings with pronounced changes in cross-section brought about by ribs or lugs, for example, are best drop-forged, whilst other items with little surface configuration are more suitable for press forging. Forging machines are used for producing upset and pierced circular-section items with long shanks.

The metals suitable for drop forging include not only steel, but also aluminium, copper, zinc, and brass, as well as alloys of these metals.

Position of parting plane.—The problem of where to place the parting

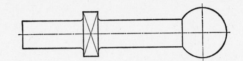

Fig. 184.—Drop forging with single axis of symmetry

plane in drop-forging dies is discussed below with the aid of some very simple examples, and examined in the light of experience.

If the work has only one axis of symmetry (fig. 184) then the parting plane should run through this axis.

If, however, the component has a plane of symmetry (fig. 185) like the lever illustrated, then there are no less than five possible die arrangements, each of which must be examined for suitability.

The most obvious solution is to take the plane of symmetry (fig. 186*a*)

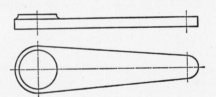

Fig. 185.—Drop forging with single plane of symmetry

as the parting plane. This, however, would present serious difficulties. Deep die impressions would be needed, and the lever would have to be pre-forged almost to the finished condition to permit insertion into the impression. The aim therefore should be to place the impression in the die in such a way that it is as flat as possible, and this is a principle that the designer must note. The layouts shown in *b*, *c*, *d*, and *e* are therefore possible. Of these, *c* is the simplest and cheapest. Method *d* calls for radiused edges, whilst *e* would entail placing the lever in the upper die, which would be pointless.

If, on the other hand, it were proposed to forge the lever-type handle (fig. 187) by placing the parting plane to give the flattest possible die impression (*b*), the reaction from the forging shop would be that it is a

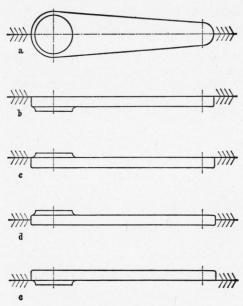

Fig. 186.—Position of forging in dies

fact confirmed by experience that method *a* allows the head to be forged more satisfactorily. The type of die impression shown in *a* would therefore be the one used, even though it is more expensive.

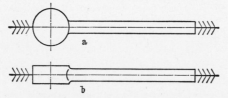

Fig. 187.—Providing for proper forging throughout the piece

(*a*) satisfactory   (*b*) unsatisfactory

Even these elementary examples are sufficient to show how advisable it is for the designer to consult the shop.

Forging draft.—Draft must be given to the sides of a forging to permit easy removal from the dies. Radiusing is sufficient on small items (fig. 188*a*), but for larger work it is necessary to provide draft as at *b*, *c*, and *d*,

Fig. 188.—Providing for easy removal from the die

the slope given to the sides of the upper die being made greater than that used for the lower die (*d*).

Forging ribs.—To obtain a satisfactory flow of metal in the dies, forgings with rib-like features are struck with the ribs uppermost, but in the manner shown in fig. 189*a*, not as at *b*.

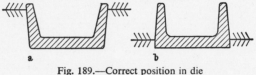

Fig. 189.—Correct position in die

(*a*) right   (*b*) wrong

Forging of hollow bodies.—The same considerations also apply to the forging of pot-shaped hollow bodies (fig. 190).

Even bodies with an inward-directed rim can be forged (fig. 191).

Fig. 190   Fig. 191

Size variations.—Permissible size variations depend on the dimensions of the forging. It is often preferable to know the weight variation rather than the size variation. For a forging weighing 10 kg, for example, the departure from the specified weight should be only 4 per cent.

Some examples are given below to show the designer the kinds of shapes which can be drop-forged.

Offset shapes: fig. 192
Angular shapes: fig. 193
Curved shapes: fig. 194
Forked shapes: fig. 195
Annular shapes: fig. 196
Ribbed shapes: fig. 197
Hollow shapes: see fig. 190 and fig. 191

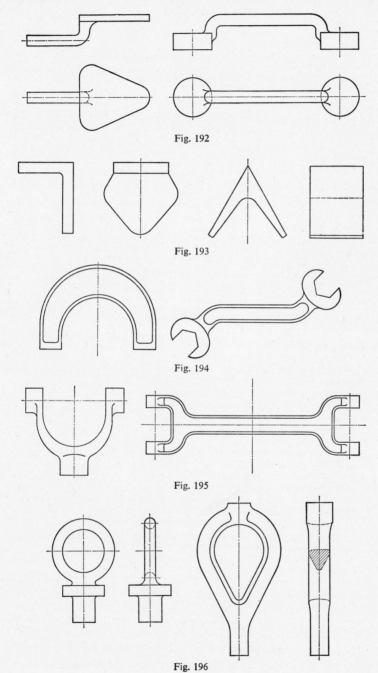

Fig. 192

Fig. 193

Fig. 194

Fig. 195

Fig. 196

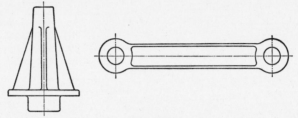

Fig. 197

Even items containing holes which would take a lot of time and much sacrifice of metal if produced by machining can be made economically in a forging machine without entailing a lot of finishing work (fig. 198).

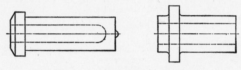

Fig. 198

Rules

1. First make a check to see whether the work could be produced more easily by welding, drawing, or stamping.
2. Position the parting plane so that the impression in the die is as flat as possible.
3. Design forgings so that a level smooth seam is obtained.
4. Avoid high ribs and deep sinkings in the dies.
5. Give proper draft to the side walls.
6. Provide the right conditions for satisfactory metal flow.
7. Avoid multi-part dies wherever possible.
8. Avoid undercuts.
9. Avoid sharp edges owing to the high rate of die wear which they cause.
10. Avoid machined surfaces and unnecessary machining as far as possible.
11. Provide machining allowances to B.S. 1718: 1951.
12. Avoid re-designing forgings because this usually renders the dies useless.
13. Note the permissible size variations laid down in B.S. 1718: 1951.
14. Provide the forging shop with a drawing of the component containing all necessary information. Apart from giving all the important dimensions of the forging, this drawing should also give information on the following points:
    (*a*) Parts to be machined.
    (*b*) Clamping surfaces, if any.
    (*c*) Dimensional accuracy required.
    (*d*) All principal cross-sectional shapes.
    (*e*) Outline of the finished forging indicated by chain line.
    (*f*) Correct position in die together with indication of flash line.

Casting, forging, or welding?—Before he decides to make a given part by forging, the designer must ask himself whether in the circumstances it would be more economical to produce it by casting or welding.

The question whether to cast, forge, or weld is not an easy one to answer. As a general rule it can be stated that parts are only forged or welded when small quantities are involved. But where is the boundary line? In this situation only an accurate calculation can decide the issue.

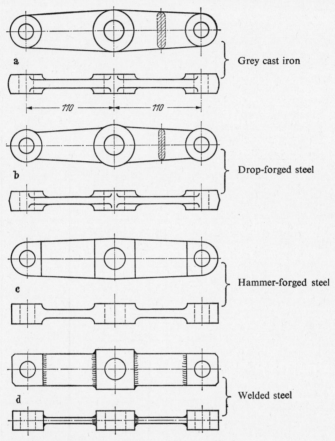

Fig. 199.—Equivalent links made by different methods

As an example, consider an ordinary link. A part like this can be made equally well as a casting, a drop forging, a hammer forging, or a welded fabrication. Fig. 199 illustrates how the design varies to suit the method of manufacture used. Now the designer must decide on the cheapest method of manufacture, and the answer to this question will depend on the quantity required. If the choice is confined to welding or forging, then it can be said right away without any calculation that the welded item will be the cheaper, and this indeed is true of whatever quantity is involved, since, in contrast with casting and drop forging, the quantity required does not influence unit price in so far as the first-mentioned methods of manufacture are concerned.

The calculated unit costs for the various methods of manufacture are detailed below (Table 3).

TABLE 3

Cost of component	Casting	Drop forging	Hammer forging	Welding
Link (fig. 199)				
Material cost per item	1s. 1d.	1s. 1d.	1s. 4d.	10d.
Pattern (die) cost	£12. 18s.	£97	—	—
Labour cost for blank	1s. 1d.	1s. 7d.	14s. 3d.	6s. 1d.
Labour cost for machining (per item)	9s. 1d.	6s. 2d.	9s. 4d.	7s. 10d.

The graphical representation of the cost of manufacture by the various methods shown as a function of quantity (fig. 200) clearly demonstrates

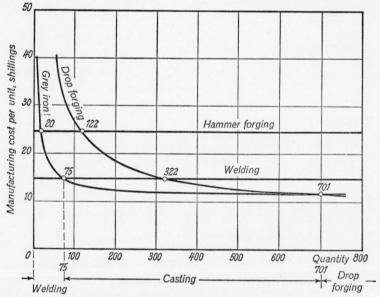

Fig. 200.—How the quantity required influences choice of manufacturing method

the areas in which the various production methods can be economically employed. It is interesting to note in this connection that for quantities of twenty and upwards casting is already cheaper than hammer forging, and that for quantities of 120 and upwards drop forging becomes more economical than hand forging despite the high cost of the dies.

(i) Designing for manufacture by machining methods

During instruction in machine drawing it will have been pointed out to the beginner in connection with the dimensioning of drawings that all dimensions and machining allowances of significance in the manufacturing process must be indicated in the drawing. The production shops must not be allowed to make even the smallest deviation from the specified dimensions and allowances.

The very heavy burden of responsibility which this places on the designer is a fact that cannot be emphasized too early. The designer can only fulfil this task if, as mentioned previously, he is familiar with production methods and is able during the designing of the part to visualize its machining from the blank right through to the finished product.

A design always starts from a simplified basic form which follows the lines of the basic design. If the part concerned is a lever, for example, then the shape given to it will depend very much on whether it is to be made by casting or forging with subsequent machining. If forging is the only manufacturing method considered, then the designer must be able to visualize the shape which the lever takes on during forging when the blank is set down, drawn, pierced, etc., under the hammer. The further designing of the part can then proceed under constant checking for maximum economy of manufacture.

Any attempt during design exercises to make corrections to a design with a view to making it less complicated and less uneconomical to manufacture usually call forth from the beginner the objection that, after all, the part could have been manufactured as originally designed. And so indeed it could, but at what cost in terms of time, material, or expensive fixtures? One must always be mindful of the fact that for economic reasons alone, every design must be capable of being manufactured as cheaply as possible. It follows therefore that every design which does not satisfy this condition is a bad one, or rather a wrong one.

There is a further important factor which deserves special mention. In the past, a great deal of care was devoted to ensuring that all the components of a machine were machined all over, so that everything literally flashed and gleamed and conveyed an impression of intrinsic worth. This of course entailed a very considerable amount of extra work and expense for no useful result. In this connection the modern attitude is the correct one, namely that machining is only justifiable when required to minimize sliding or rolling friction or to give greater strength. A lever, for example, will have only its bores and sides machined.

It is a long-established fact that the student designer always falls into the same errors in his design work. For this reason certain firms have catalogued the errors made and issued them as " Hints for the Designer ". But examples of this kind can be added to without limit and, since the beginner cannot be expected to memorize them all, the obvious approach is surely to examine the basic factors which are constantly offended against in the context of machining. Such an examination reveals that the causes of faulty designing can be traced back to just a few factors.

Design errors can generally be avoided if the following points are taken into account:

Machinability	Avoidance of redundant fits
Economy	Accessibility
Clampability	Ease of assembly
Existing tool equipment	

Designing for machinability.·—A striking example of wrong design is given in fig. 201a which illustrates a rod end required in steel with all-over

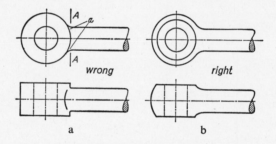

Fig. 201.—Designing for machinability

machined finish. It will be seen that the shank can be turned as far as cross-section A, and the sides of the shank and outside of the eye milled. There still remains a small area at *a*, however, which can only be reached by hand filing. The correct design for economical machining is shown in fig. 201b.

The newcomer to designing very often forgets that a cutting tool needs a certain amount of run-out or over-travel if it is to impart a clean finish. This applies equally to turning, planing, milling, grinding, and reaming. For examples see fig. 202.

By appropriate design measures the designer can often simplify machining operations. It is more difficult to cut grooves in the bush of fig. 203 than in the rod. For the same reason the recess will be made in the eccentric and not in the strap.

While a drill is cutting it should always meet equal resistance on its

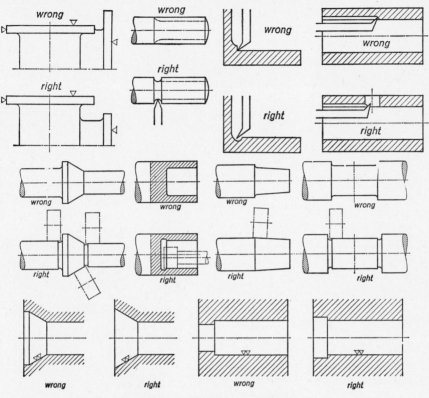

Fig. 202.—Designing for machinability

cutting edges. This condition will only arise if the entry and exit surfaces encountered by the drill while it is at work are perpendicular to its central axis (fig. 204c). In the situations shown in fig. 204a and b the drill will be

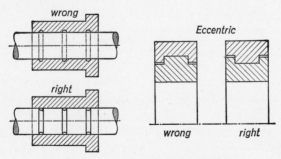

Fig. 203.—Examples of work simplification

deflected to the side. On no account should the drilling of sloping surfaces be attempted (fig. 204*a*).

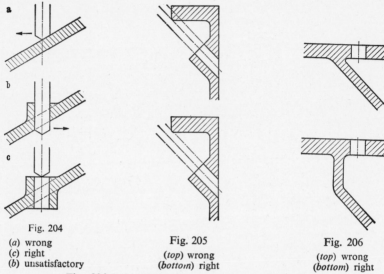

Fig. 204
(*a*) wrong
(*c*) right
(*b*) unsatisfactory

Fig. 205
(*top*) wrong
(*bottom*) right

Fig. 206
(*top*) wrong
(*bottom*) right

Figs. 204–206.—Right and wrong positions for drilled holes

Holes should not be placed too near the edge of the work, since the metal may break away if it is cast iron; on the other hand, if it is steel it will tend to deflect at the thin cross-section while drilling is in progress and will spring back afterwards. The result will be a hole which is out-of-round.

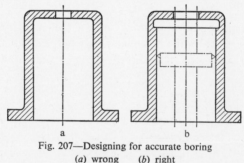

a   b
Fig. 207—Designing for accurate boring
(*a*) wrong   (*b*) right

Thoughtlessness on the part of the beginner may lead to holes being so positioned that they are inaccessible to the drill (fig. 205). Situations like those shown in fig. 206 endanger the drill and should be avoided. Parts intended to be machined in a boring mill must have adequate openings to admit the spindle (fig. 207). Accurate boring cannot be expected from a

spindle without any outboard support, and provision must therefore be
made for an outer bearing (fig. 208).

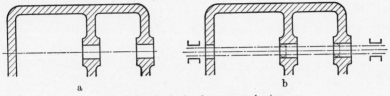

Fig. 208.—Designing for accurate boring
(*a*) wrong   (*b*) right

Designing for economy.—When one speaks of economy in connection
with machining one is of course thinking of those costs which arise exclusively through the manufacturing process. Economic designing, however, depends also on other important factors as we have already seen on
page 25 (fig. 3). The designer must consider the economic aspect of every
problem right from the start, that is to say even while he is working out
the best basic design. Economic considerations guide his search for the

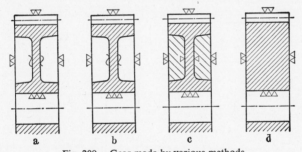

Fig. 209.—Gear made by various methods

(*a*) Casting   (*b*) Drop forging   (*c*) Machined from solid   (*d*) as (*c*) but left solid

most suitable material, whilst the vital importance of keeping manufacturing costs down entirely dominates the designer's work when he is deciding
on the shape to give to the component.

The cost question, that is to say the economic aspect, influences either
directly or indirectly practically all design and, of course, production; a
final decision on all such matters can be made only after calculation. In
small businesses the designer usually has to carry out the costing in agreement with the works; larger firms have special costing departments. The
importance of costing as a means of deciding between various manufacturing methods can be illustrated by a simple example.

A gear (fig. 209) will perform equally well whether made by methods *a, b, c,* or *d.* If the choice is limited to *c* and *d,* it is possible to say right away without any calculation of costs that version *d* will always be cheaper than *c,* since less metal removal is needed. When larger quantities are involved, however, the stage will certainly be reached where version *d* will become dearer than either *a* or *b,* despite its simplicity from the manufacturing and machining viewpoints.

A decision as to which method to use can only be taken by listing the component costs in detail as shown in Table 4.

TABLE 4

Cost of component	Version *a*	Version *b*	Version *c*	Version *d*
Gear (fig. 209)				
Material cost per item	11*s*. 4*d*.	9*s*. 11*d*.	14*s*. 8*d*.	15*s*. 11*d*.
Pattern (die) cost	£12.7*s*.6*d*.	£134. 15*s*.	—	—
Labour cost for blank	10*s*. 10*d*.	3*s*. 4*d*.	7*s*. 11*d*.	—
Labour cost for machining (per item)	30*s*. 4*d*.	31*s*. 7*d*.	63*s*. 5*d*.	37*s*. 10*d*.

By using these figures it is possible to show how the choice of manufacturing method is influenced by the quantity required (fig. 210).

As mentioned previously, the cost analysis shows that version *c* is the most uneconomical one and incapable of competing with the other versions. For any quantity up to 303 the lowest manufacturing cost per

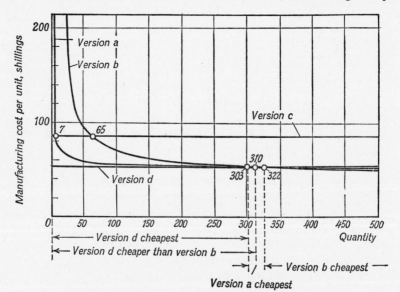

Fig. 210.—How quantity affects the economics of manufacture

unit is yielded by version *d*. Only over the short range from 303 to 322 units is the cast-steel version cheapest. From 322 units upwards manufacture by drop forging is cheapest.

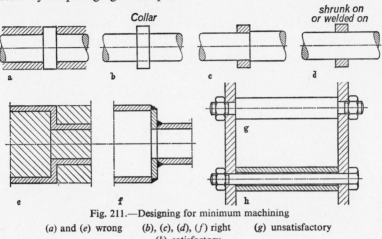

Fig. 211.—Designing for minimum machining

(*a*) and (*e*) wrong   (*b*), (*c*), (*d*), (*f*) right   (*g*) unsatisfactory

(*h*) satisfactory

Fortunately the designer is often able without making a cost analysis to design into his product the necessary conditions for economical manufacture. A few simple examples will demonstrate this.

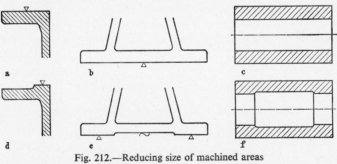

Fig. 212.—Reducing size of machined areas

(*a*), (*b*), (*c*) unsatisfactory   (*d*), (*e*), (*f*), satisfactory

Low-cost manufacture requires that machining work be kept to the minimum. The designer meets this requirement by appropriate form-design measures. Compare, for example, fig. 211*a*, *e*, and *g* with fig. 211*b*, *d*, *f*, and *h*.

The cost of manufacture can also be lowered by reducing the area of machined surfaces. Compare, for example, fig. 212*a*, *b*, and *c* with fig. 212*d*, *e*, and *f*.

Rechucking or reclamping of the work and resetting of tools take up time and therefore add to the cost of machining. Compare the examples of incorrect and correct form-design shown in fig. 213.

The manufacturing method adopted often by itself brings about a considerable cheapening, e.g. by substituting welding for forging, forging for casting or vice versa, welding for casting, welding for riveting, etc.

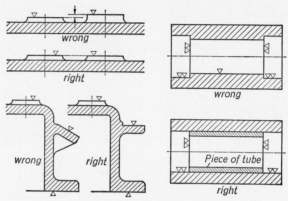

Fig. 213.—Avoiding extra work

A change to a different type of material, e.g. the use of steel or cast steel or even aluminium instead of cast iron, can also bring economic advantages.

Page 168 lists the points which enable an economical design to be achieved without the designer first having to make a cost analysis.

Designing for clampability.—Provision for secure clamping of the work is the essential condition for accurate machining. This must be taken into account during the form designing of the work, so that the necessary location faces, bosses, lugs, etc., can be provided (fig. 214).

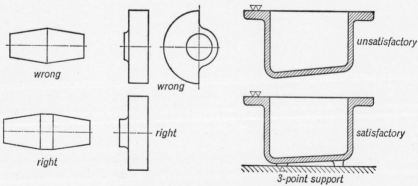

Fig. 214.—Designing for clampability

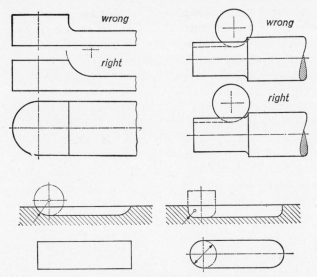

Fig. 215.—Designing for existing tool equipment

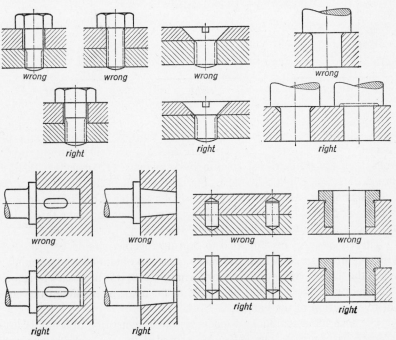

Fig. 216.—Avoiding redundant fits

Designing for existing tool equipment.—The designer's policy should be to design in such a way that special tools are not needed. Every designer ought therefore to be in possession of a list of available tool equipment, so that he can specify dimensions for machining operations correctly and avoid using non-standard sizes in his drawings (see examples, fig. 215).

Designing to avoid redundant fits.—The duplication of fits should always be avoided, since such fits can only be achieved, and then only as an approximation, by using very close tolerances entailing extra work and high cost; furthermore, they usually have the effect of defeating the purpose for which they are intended (fig. 216).

The examples illustrated represent errors that are continually made by beginners in this context.

Designing for accessibility.—This is a principle which is regularly offended against when a design is produced without due regard for the

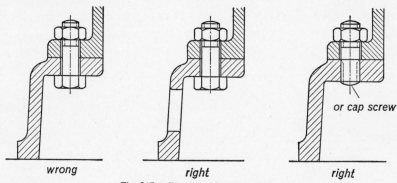

Fig. 217.—Designing for accessibility

whole picture from manufacture through assembly to convenience of operation, etc. Errors of this kind are specially painful to the beginner when they are pointed out to him by someone else, since they betray thoughtless designing on his part. Fig. 217 will illustrate this.

Designing for ease of assembly.—Ease of assembly is one of the criteria of the suitability of a design. The designer must therefore take this factor into account in his work (figs. 218 to 221).

Turned components intended to bolt up together should be provided with a centring spigot which will make it easy to obtain coaxial assembly. When the flanges are not of circular form (fig. 218) it is necessary to resort to dowel pins; these should be spaced as far apart as possible to ensure positive location of the flanges.

The accurate assembling of machine frames consisting of a number of separate parts presents no difficulty, provided that suitable tenons and

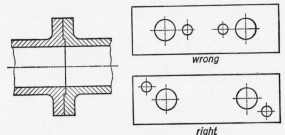

wrong

right
Fig. 218.—Designing for ease of of assembly

locating faces are furnished; in the absence of such features, however, tedious aligning and re-measurement must be carried out before the individual parts can be dowelled together (fig. 219).

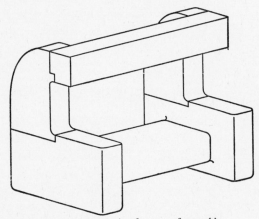

Fig. 219.—Designing for ease of assembly

Speedy assembling and dismantling call for easy access to all parts. To facilitate dismantling, holes are provided to take jacking and withdrawal screws (fig. 220).

It is often possible to lighten the task of assembly by quite simple measures. For example, the insertion of a pin into a hole may prove difficult unless the edges concerned are bevelled (fig. 221b).

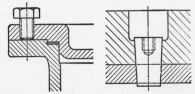

Fig. 220.—Designing for ease of assembly

Long bushes are often stepped on their outside diameter so that their effective length for fitting purposes is shortened. It would be bad practice, however, to make the shoulder exactly in the middle because this would mean that two edges of the bush would have to be lined up simultaneously with their receiving diameters. This would be quite a difficult task as will be seen by comparing fig. 221c with d.

On the other hand, if the shoulders are so positioned that the leading portion of the bush enters its bore first, then the insertion of the remainder will present no difficulty.

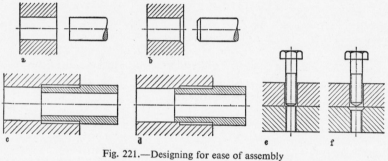

Fig. 221.—Designing for ease of assembly

(a) and (c) wrong   (b) and (d) right   (e) unsatisfactory   (f) satisfactory

The threading of long screws into deep holes is not easy. By pointing the screw and countersinking the tapped hole, however, the designer can take the element of tedious hit-and-miss out of the operation (fig. 221f).

Correct designing can eliminate the risk of wrong assembly by

1. Using different pin diameters.
2. Marking gears.
3. Specifying the setting of adjustable linkages.
4. Providing unsymmetrical holes for detachable cranks.
5. Using unequal hole pitches, numbering, or other forms of marking for circular covers.

The best way to check for convenience of assembly during the drawing-office stage is to let another designer enter the details in an assembly drawing.

5. HOW THE SPACE FACTOR INFLUENCES FORM DESIGN

During their first design exercises all students discover that not every item can be designed independently without allowance for other items which it affects. Spatial limitations thus apply even when the customer refrains from expressing any requirements in this connection. The overcoming of these intrinsic limitations can by itself cause the designer a great deal of difficulty.

To this must be added the fact that economic considerations oblige the designer to save material, that is to say to design in a way which will keep the size as small as possible. This requirement likewise imposes restrictions on the shape of the object designed.

It can be said therefore that every design assignment involves the designer in a struggle with complications introduced by the space factor. This applies with special force when space restrictions are imposed right from the beginning as, for example, when a structure must be kept within a specified size or fitted into a given space.

There are certain areas of mechanical engineering, such as automobile and aircraft construction, where requirements of this kind have to be constantly met by the designer. A stationary boiler can be evolved with complete freedom as far as its occupancy of space is concerned. With a locomotive boiler this cannot be done, since the width is limited by the loading gauge, so that there is a definite restriction on the amount of development in width which can be pursued. The designer required to develop a steam boiler for automotive duty is faced with even greater difficulties. A boiler of this kind must be housed in the smallest possible space, and this presents a problem which can only be solved by abandoning conventional steam-boiler design and adopting a tubular boiler with forced flow and oil firing.

The designer cannot be expected to meet unreasonable space requirements. There is a minimum size limit to the development of every machine component and every structure; there is a limit to what is practicable. A surface condenser of given capacity operating under optimum conditions requires a certain minimum space. The only variables are the diameter and the length; alternatively, the installation can be split up into a number of smaller units. The position is much the same with electric motors. If the diameter has to be reduced for reasons of space, then the only way to maintain the same power output and speed is by making the motor longer.

There are several tricks of the trade which can be used by the designer seeking to cut down the size of object being designed. Dimensions can be made smaller by using a material offering higher strength. This is a method which is put to good use in lightweight construction as we shall see later. Space can also be saved by switching from cast or riveted construction to welded construction. Admittedly the gain is not very large, but after all the saving of even a few millimetres can be of vital importance in certain situations.

There are various possible solutions to every engineering problem. As we have seen earlier, the designer selects from the best-suited solutions on the basis of the requirements raised by the particular problem, and one of these is the conditions imposed by the space factor. A simple example will illustrate this.

A 1:20 reduction ratio can be obtained by using a two-stage spur-gear

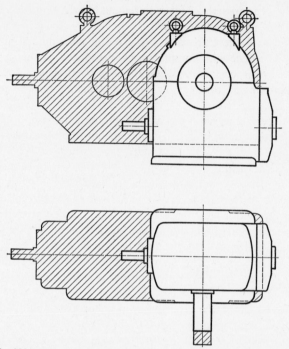

Fig. 222.—Two equivalent gear reducers (courtesy: Flender-Bocholt)

reducer or a worm reducer. Normally it would be decided to use the former, because the latter would probably be more expensive and would absorb more driving power owing to its lower efficiency. The worm reducer, however, takes up very much less space than a spur-gear unit. If, as is true of the Cavex transmission, the sliding conditions become so favourable as a result of the use of a hollow-flank worm that efficiencies virtually equal to those of spur gearing are obtained, then the designer need have no hesitation in choosing the space-saving Cavex system (fig. 222).

A few examples of wagon tipping gears actually in use (fig. 223) will show how varied are the space requirements which can arise in the solving

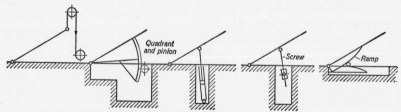

Fig. 223.—Space requirements of different forms of wagon tipping gear

of a problem. From the large number of practicable kinematic solutions available the designer will usually find one suited to his purpose.

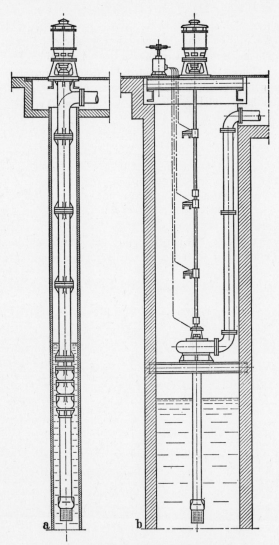

Fig. 224.—Two different types of deep-well pump
(courtesy: Klein, Schanzlin & Becker)

Deep-well centrifugal pumps (fig. 224) provide a clear example of the way in which the organic structure of a piece of equipment can be varied to adapt it closely to local conditions. In fig. 224*a* the driving

shaft runs through the riser pipe, and the diameter of the centrifugal
pump is so reduced by multi-stage construction that this unit, unlike
the one shown in fig. 224*b*, can be installed in a well of quite small
diameter.

An effective way to secure a reduction in size consists in making a
satisfactory choice among the various factors determining the action of the
equipment or device concerned. If, for example, the problem is to build
a heat exchanger of maximum compactness, the designer will know that
the quantity of heat exchanged is given by the general equation

$$Q = \alpha.F.\Delta t.z$$

where

        α = heat-transfer coefficient in kcal/m h deg C,
        F = area of heat exchanging surface in m²,
        Δt = temperature difference in deg C,
        z = time in hours.

Since the size of the unit is principally determined by the area F, the
designer will endeavour to make the other factors α and Δt as large as

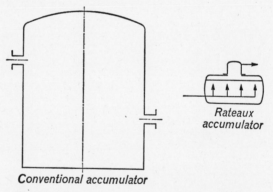

Rateaux
accumulator

Conventional accumulator

Fig. 225.—Two steam accumulators of equal capacity

possible. Among many factors which affect the heat-transfer coefficient,
three which influence it very strongly are the velocity of the cooling
medium, the prevailing flow conditions (laminar, turbulent, parallel-flow,
or counter-flow), and the condition of the surface. The designer thus has
many opportunities of keeping the size of the unit small by cutting down
F through choosing suitable values for the other factors.

To obtain an extreme example of this it is only necessary to compare a
conventional steam accumulator with the Rateaux accumulator (fig. 225).

Whereas, for example, for 1000 kg of steam the conventional accumulator requires a capacity of 1722 m³, the Rateaux accumulator requires a capacity of only 57 m³. The volume ratio is therefore

$$\frac{\text{volume of Rateaux accumulator}}{\text{volume of conventional accumulator}} = \frac{1}{30}$$

The space factor determines design to a considerable extent. The designer, however, has various means available for satisfying the requirements.

1. In the first place he will endeavour, by varying the form of the unit he is designing or changing its layout, to adapt it to the space conditions specified for it.
2. Other aids serving this end are subdivision into smaller units, the use of high-grade materials, and the adoption of welding.
3. If these devices fail to give the desired result, the designer is obliged to resort to a new working principle.
4. A satisfactory choice from among the possible methods of carrying out the required function can lead to a suitable solution.

The designer needs to have had a certain amount of practice in fitting in with given space limitations. For the beginner there is a simple method of acquiring confidence in this aspect of design at an early stage. He should

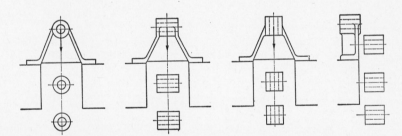

Fig. 226.—Design problem involving different space limitations

never content himself with the solving of a design problem, but should always ponder how the design would look if a change were made in the space available.

Even for a straightforward bearing bracket the space limitations can be varied within wide limits. By systematically modifying the space limitations and adapting the design to the resulting new conditions it is possible even for the beginner to gather design experience and acquire skill in adaptation.

Exercise problems

Problem 27

It is proposed that a bin lid shall be capable of being opened by hand until it is in the vertical position. The simplest method would be to use a counterweight as indicated in fig. 227. In the present layout, however, this is not possible because the bin stands very near the wall.

In what way can the weight of the lid be exactly balanced in every position? (Basic design is sufficient.)

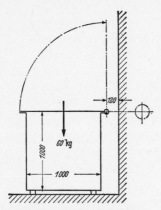

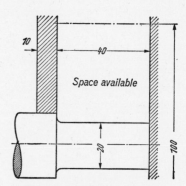

Fig. 227.—Dimension diagram   Fig. 228.—Dimension diagram

Problem 28

A V-belt pulley taking two V-belts (13 × 8 mm) and having a diameter of 100 mm must be mounted as close as possible to a housing wall and carried on rolling bearings. The dimensions to be observed are given in fig. 228.

6. HOW SIZE INFLUENCES FORM DESIGN

When presented with a new design assignment, every designer, and the beginner in particular, starts by looking for a usable existing design. For the practising designer this is usually a very simple matter, since there is always a whole range of proven designs on record. The student, however, is not in this fortunate position. For the most part he only has access to illustrations in books. By taking over model designs discovered in this way, however, the student designer runs a big risk, because he knows nothing of the size and loading factors involved. Unfortunately it is no longer customary to give the scale alongside illustrations. It may therefore happen that the model taken over by the student is a design intended for relatively large forces and dimensions, whereas the problem with which he is concerned involves relatively small dimensions and forces. The result,

drawn to the same scale might be on the lines illustrated (fig. 229). It will
be seen that the bracket has taken on a complicated shape.

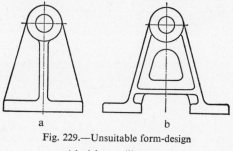

Fig. 229.—Unsuitable form-design

(*a*) right   (*b*) wrong

The way size influences form can be appreciated merely by considering
an ordinary bearing bracket (fig. 230). The larger the principal dimension
H becomes, the lighter the type of form used. If type *a* were used as a

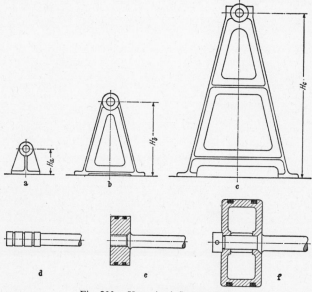

Fig. 230.—How size influences form

model for *c*, for example, the result would be an impossibly heavy shape.

Similar considerations apply to piston design. For a piston of 10 mm
diameter all that is needed is a ground stub without rings but possibly with
grooves turned in it (fig. 230*d*). Up to 80 mm diameter solid construction

7

can still be used (fig. 230e) but for still larger diameters cavity construction will be adopted (fig. 230f).

Caution and an appreciation of size and loading conditions are therefore needed when using existing designs as patterns.

7. HOW WEIGHT INFLUENCES FORM DESIGN (LIGHTWEIGHT CONSTRUCTION)

It is the endeavour of every designer, whether he is concerned with the design of individual components or of complete machines, to save material by making the fullest possible use of it, and to avoid over-dimensioning, even though nothing is laid down regarding weight.

Indeed, so much importance has been attached to saving material and to building for lightness that it has been laid down (by Kesselring) as a fundamental principle of design activity that:

Design development should be directed so as to cut material cost to the minimum.

There is no doubt that manufacturing costs are kept down by observance of this principle which also confers other advantages to be considered later. There are some branches of mechanical engineering—automobile and aircraft engineering, for example—in which the demand for lightness obliges the designer to seek special ways and means of providing it.

The various principles used to secure maximum saving of weight can be grouped under the designation " lightweight construction ". Such are the benefits conferred by lightweight construction that other branches of the mechanical-engineering industry have made good use of them even when, as with machine tool building, for example, there is no urgent need to save weight.

There are of course plenty of publications dealing with lightweight construction, but they only treat individual aspects of it. It seems appropriate at this stage therefore to give below a summary of all the factors which contribute to success in the field of lightweight construction, so that the student designer can familiarize himself with them.

The factors concerned are as follows:

1. Optimization of form.
2. Best possible estimate of strength.
3. Use of welding instead of riveting.
4. Use of welding instead of casting.
5. Use of high-grade steels.
6. Use of lightweight materials.
7. Use of special sections.

8. Use of novel mechanical elements.
9. Saving of weight through basic change of layout.
10. Optimum choice of factors which determine the mode of action.

(a) Optimization of form (lightweight construction)

Lightweight construction means taking material away from wherever it is not being fully utilized without thereby impairing the strength of the structure. The ideal postulated for lightweight construction is therefore a body offering equal strength. The primary condition to be observed in any mode of lightweight construction is appropriateness of form design. This then forms the basis on which the individual dimensions can be calculated for maximum fidelity to the facts.

What then are the factors which the designer must bear in mind if he is to achieve optimum form? The central fact to be kept in view all the time is that the part to be designed should represent the closest possible approach to the ideal solution of the problem in hand. The designer must therefore ensure that the static, dynamic, and kinematic functions of every part to be designed are clearly understood. Even during the early stages, while the basic design is being evolved, it is essential to avoid any obscurity and small concessions, otherwise these will inevitably result in additional stressing in the form of, say, bending and torsion, and will prevent full utilization of the material.

Every engineering component is acted on by external forces which set up internal forces (stresses) in the component. The external and internal forces are in equilibrium. If the external forces are only of a tensile or compressive nature, it is an easy matter to tell how the internal forces are disposed. They must in fact run in straight lines in a direction parallel to the external forces. Therefore, if he is to save material, the designer must give the object he is designing a shape that will ensure the presence of material in the direction of the compressive or tensile load to take the internal forces set up. If the external or compressive forces give rise to a force pattern which is not of the straight-line type, but shows single or multiple curvature, this can always be taken as an indication that bending stresses are present in addition to tensile and compressive stresses. Configurations of this kind always entail a substantially larger amount of material and are therefore unsuitable for lightweight construction. The examples illustrated (fig. 231) bring this out clearly without any further explanation.

Special attention must also be paid to points where the external forces are applied, and also to joints, since it is in these areas—disregarding welded construction—that considerable complication may be introduced into the force transmission scheme and consequently into the force pattern

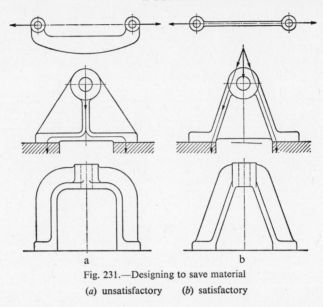

Fig. 231.—Designing to save material

(*a*) unsatisfactory   (*b*) satisfactory

(fig. 232). At such points therefore considerable extra bending stresses are liable to occur and to give rise to stress peaks which call for care in form design.

Whenever bending stress is involved, even if a plain straight body is concerned, the path taken by the line of principal stress already tends to

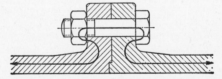

Fig. 232.—Force pattern obtained at a bolted fastening

become quite complicated. A still more complex situation arises when a torsional force is superimposed on bending. Under these conditions the lines of principal stress follow three-dimensionally curved paths.

The student designer must therefore be advised as follows: when tensile and compressive loading exists, try by suitable form design to avoid superimposing any bending effect, and make sure that a straightforward force pattern is obtained, that is to say a pattern consisting of gently curved paths.

Attention must also be paid to questions of fixing and support. Every structural component is joined in some way to another. Appropriate choice of supporting and fixing methods will allow weight to be saved in this direction also. If bending loads are primarily involved, then load capacities and deflections will vary on the lines indicated in the diagram (fig. 233) assuming the same cross-sectional size throughout.

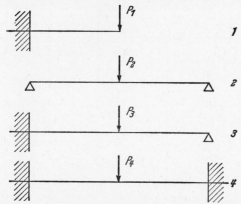

Fig. 233.—Beams yielding different load-carrying capacities and deflections for the same cross-section

The relative strength for these systems is as follows:

for systems supported or fixed as in	1	2	3	4
the load capacity is	1	2	2·7	4
and the deflection	1	$\frac{1}{2}$	1/4·5	$\frac{1}{8}$

configuration 1 being taken as unity.

An illustration of the way in which the designer can save material by providing suitable supporting and fixing arrangements for a structure is given by the following simple example (fig. 234). It is assumed that a projecting roof is to be built in steel for maximum lightness and uniform loading.

What are the available designs and which one promises to be the lightest in weight? The designer has eight types to choose from. Examples of all the types shown are to be found in actual practice. If a particularly lightweight mode of construction is required, however, the optimum solution will be immediately found on examination. The self-supporting type 1 will certainly be the heaviest owing to the fairly large bending moment at the fixing point. A rather more satisfactory arrangement is beam 2 which is shaped to yield a member of uniform strength. The beam can also be supported by an upright (type 3), but even this type can be made still lighter. Type 4 uses compression members which, however, result in larger size and heavier weight than the system using tension members (types 5 and 6). Of these two latter versions, 6 is probably the more satisfactory with regard to weight saving. The only further point to investigate would be the angle α at which the tension members are inclined. Small angles yield large tensile forces but short lengths, whereas large angles yield smaller forces but correspondingly greater lengths and weights. Calculation gives 45° as the most satisfactory angle. It would also be necessary to check whether lattice girders could compete with type 6. Of the last two arrange-

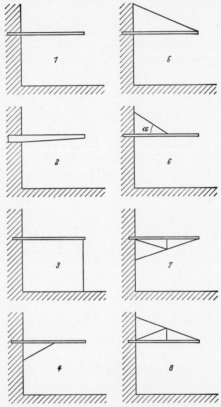

Fig. 234.—Various methods of supporting a projecting roof

ments illustrated, type 8 yields the lighter structure owing to the fact that fewer compression members are needed than for type 7. The designer must also deal with similar problems, such as arrangement of members and choice of sections, when designing lattice type structures.

Maximum utilization of material leads to the idea of taking it away from any place where it is not being used to the full for transmitting the forces involved. This is the position with those layers of material adjacent to the neutral plane. If material is taken away from this point the forms obtained are the familiar I, channel, and hollow sections, which are the most satisfactory types of cross-section for the purposes of lightweight construction.

When bending loads arise, the most satisfactory types of cross-section for cast iron are the unsymmetrical shapes ⊥, ⊔ owing to the big difference between tensile and compressive strength, whilst for steel and cast steel the most suitable forms are the symmetrical ones I, ⊏, ⊡. A rectangular section standing on edge withstands bending loads quite satisfactorily, but a comparison of this type of section with the I-section having the same section modulus (fig. 235) reveals that by choosing the latter a weight saving of 47 per cent can be achieved.

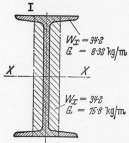

Fig. 235.—Advantage offered by I-section compared with rectangular section having same section modulus.

Fig. 236 (according to Kloth) shows how material utilization can be carried to still farther limits in comparison with a rectangular cross-section.

Form of beam	Amount of material in kg	in %
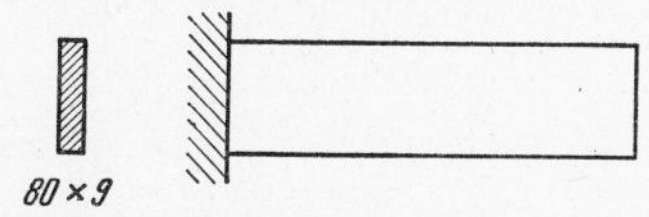 80 × 9	5·9	100
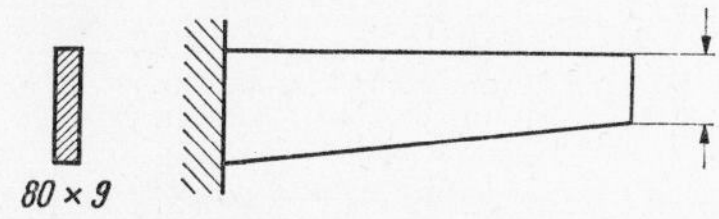 80 × 9	4·4	74·6
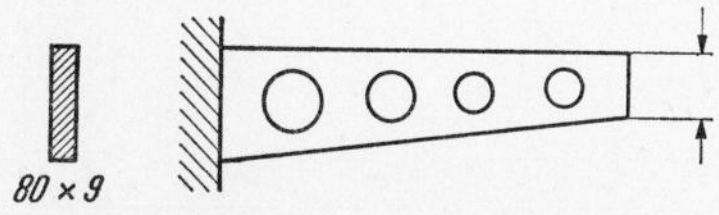 80 × 9	4·0	67·8
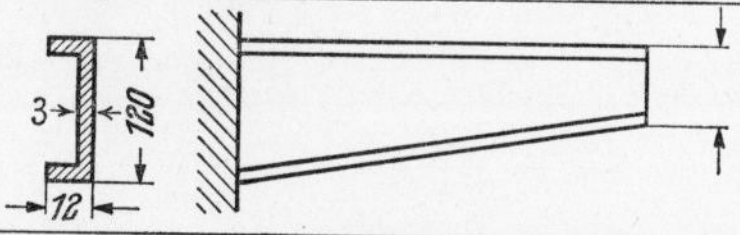 3→ 120, 12	2·5	42·4
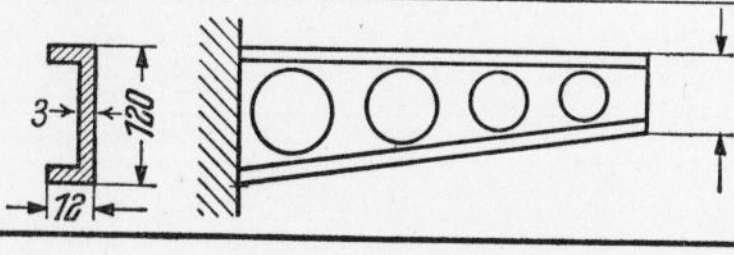 3→ 120, 12	1·7	28·8

Fig. 236.—Weights of different types of beam carrying the same load (according to Kloth)

In the building of machines there are many instances where the criterion is deformation rather than strength. For nearly all machine tools the primary requirement is stiffness. The latter increases in inverse proportion to the deflection. The relationship is expressed as follows:

$$\text{stiffness} = \frac{\text{load}}{\text{deflection}}$$

$$= \frac{P}{f} \quad (\text{Kg}/\mu)$$

where f = deflection in $\frac{1}{1000}$ mm = $[1\mu]$

It is no doubt true that a body intended for static loading can be designed to offer greater stiffness to bending loads, that is greater rigidity, by being made thicker and deeper. If this is done, the moment of inertia can only be increased at the cost of a comparatively large addition to the weight. The same objective can, however, be reached by a much simpler path, namely by choosing a more favourable cross-sectional form (fig. 237). By this means the component can be made not only more rigid but also substantially lighter.

This applies both to statically-loaded and dynamically-loaded elements.

	Moment of Inertia cm^4	Deflection cm	Stiffness kg/μ	Amount of material	
				kg	%
	24	0·66	0·15	13·6	100
	19·3	0·82	0·12	12·0	88
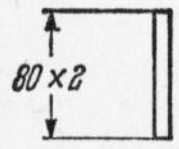	38·4	0·41	0·24	5·9	43
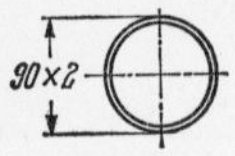	53	0·30	0·33	6·8	50
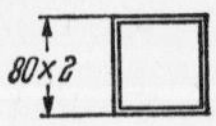	68	0·23	0·43	4·7	35
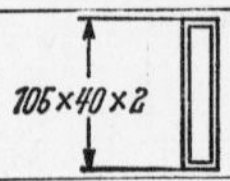	78	0·20	0·50	4·4	32

Fig. 237.—Rigidity and amounts of material used

A simple example may illustrate this. The critical speed for a solid shaft of 50 mm diameter carried in bearings spaced at 1800 mm is in the region

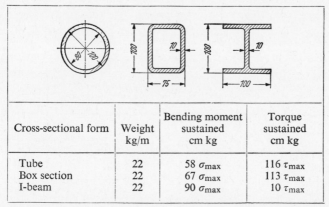

Cross-sectional form	Weight kg/m	Bending moment sustained cm kg	Torque sustained cm kg
Tube	22	58 σ_{max}	116 τ_{max}
Box section	22	67 σ_{max}	113 τ_{max}
I-beam	22	90 σ_{max}	10 τ_{max}

Fig. 238.—Bending strength and torsional strength obtained with equal amounts of material (according to Thum)

of 1900 rev/min. For a tubular shaft of 85 mm diameter and 2·5 mm wall thickness, that is with identical section modulus, the critical speed is not reached until 4100 rev/min.

We know already that a tube provides the most favourable type of cross-section to withstand torsion. Fig. 238 shows the behaviour of various cross-sectional forms under torsional loading, the amount of material being the same for each shape. It will be seen that the I-section beam can carry scarcely a tenth of the torque although on the other hand it offers the greatest resistance to bending.

By choosing a tubular cross-section a considerable saving in weight can be achieved as shown in fig. 239.

Lightweight construction in steel would be inconceivable without welding. Earlier, in the section in which designing for welding was discussed, it was pointed out that the following types of construction lend themselves to welding (fig. 155):

1. Plate construction
2. Built-up plate construction
3. Hollow construction
4. Cellular construction

Cross-sectional form	Weight
	100 %
	81·7%
	51·7%
	20 %

Fig. 239.—Difference in weight for same section modulus

The first two types of construction are only justified in situations calling for shear strength. On the other hand, when small deformations and vibration-proof structures are involved, the methods used will be hollow and, more particularly, cellular construction. These two methods offer

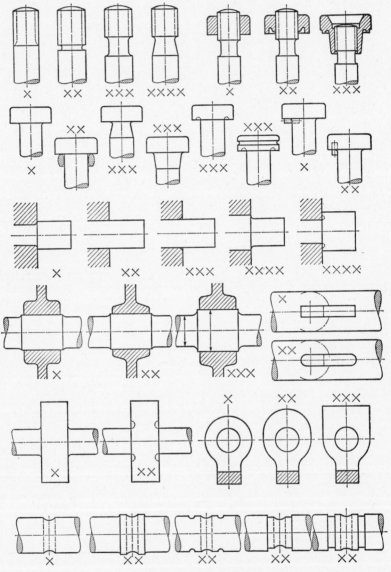

Fig. 240.—Notch effects

the greatest saving of material and maximum stiffness. They are also used extensively in machine tool building where rigidity usually plays a more important part than strength.

Measurements of stress distribution in machine components have demonstrated that stresses are non-uniformly distributed over the cross-section whenever necked elements and abrupt changes of cross-section occur, that is to say at recesses, fillets, keyways, holes, shoulders, collars, corners, edges, fins, etc. At these points there occur stress peaks which are considerably larger than the assumed uniform nominal stresses, and these peaks are a source of failure.

Many examples of the causation of notch effects and ways of minimizing them are given in the literature. The beginner can only be urged most strongly to study these publications. Some of the most common instances

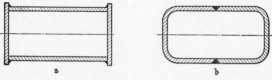

Fig. 241.—Box-type cross-sections

which the designer continually encounters are reproduced here (fig. 240). The examples indicated by a single star are the most susceptible ones, whilst those with more than one star show how unduly high stress peaks can be avoided. Reduction of the stress peaks can be brought about by appropriate form design or production measures.

It is not necessary to design welded fabrications with sharp corners (fig. 241a). By using a press brake to bend the plates it is possible to obtain cross-sections with rounded corners offering greater stiffness than is obtainable with sharp corners, plus the further advantage that fewer welds are needed (fig. 241b).

The cellular type of construction requires the smallest amount of material and at the same time offers maximum stiffness even under torsional loading. The cells are formed by thin plate of the same thickness and usually of the same size throughout, and these plates are assembled in diagonal or perpendicular arrangements between parallel walls. Fig. 242 shows some of the most usual types of arrangement.

When a structure is required to resist vibration, as may be required in many machine-tool designs, the aim should be to build as compactly as possible. The designer achieves this objective by placing the material as far away from the neutral plane as possible and by shortening the free length. This imparts greater rigidity by raising the vibration frequency.

Large flat surfaces naturally tend to flex. Advance calculation of the vibration conditions is seldom possible and the designer must therefore rely on experiment. If vibration of this kind should happen to set in at some later date, however, it is quite easy to remedy the condition in a welded structure. This is done by tying the parallel walls together by welding lengths of pipe between them. Ribs can also be used, but caution is

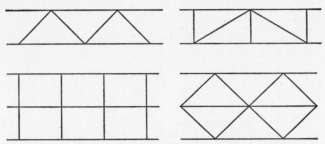

Fig. 242.—Stiffening ribs used in cellular-type construction

necessary with this method owing to the fact that deep ribs give rise to severe stress peaks. Stiffening of the walls by swaging is usually a better method. A tendency for the surface to vibrate as a whole can also be reduced by curving the walls.

The weight-saving advantages of hollow construction have been exploited by the automobile designer who makes use of the whole vehicle body, sub-frame, sides, and roof to take the loads which formerly had to be carried by the sub-frame alone. The result is a large closed tube offering

Fig. 243.—Stiffening of skin sheeting

exceptional resistance to bending and torsion. The building of the wings and fuselages of modern aircraft, too, would be unthinkable without the hollow-construction method.

Skin sheeting is stiffened by appropriate form design or by applying curvature, swaging, or corrugation as shown in fig. 243.

The removal of material from the neutral zone leads to a lightweight building element known as dual sheet which offers very high resistance to bending in all directions (fig. 244). This type of sheet has dimples pressed in at uniform spacing and these are spot-welded together. Weight savings up to 60 per cent are made possible in this way. A recent development is the use of a sandwiched layer of plastic material to hold the sheets together, and in this form the sheet is employed in the construction of self-supporting wagon bodies.

It is well known that parts subjected to compressive and buckling forces require a greater bulk of material than parts stressed only in tension. The task of the designer engaged in lightweight construction is to develop the form which will really make full use of the compressive strength of the material concerned. Since hollow sections are best suited to this type of loading, the designer must aim to secure the largest possible radius of gyration in return for relatively small wall thicknesses. Since the compressive forces and length conditions are nearly always specified, Wagner's coefficient $\sqrt{\dfrac{P}{e}}$ is a useful method for correctly determining the most satisfactory cross-sectional form.

The welded construction of large internal-combustion engines provides a very striking example of the way in which the designer can secure

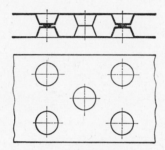

Fig. 244.—Dual sheet

a simple stress pattern which in turn permits the most economical employment of material. When castings are used for components of this kind, the lines of force take unknown paths between the cylinders and the main bearings. When through tie rods are used, however, as shown in fig. 245, the tensile forces are confined to these rods, and the remainder of the structure is relieved of load.

The outstanding successes achieved by appropriate design methods as applied to welded fabrications, and the important saving of weight obtained thereby, have led designers to apply the principles of lightweight construction to the design of castings also. After all, materials which can be cast present the important advantage of permitting an almost limitless variety of shapes to be imparted to them. For this reason it may easily happen, for example, that intricate parts made in steel (by welding) will be more expensive than corresponding parts in cast iron, although the numbers involved are small.

Admittedly the old data on cast iron, and the opinions handed down in regard to dimensioning, allowed little scope in view of the low tensile figures. Every designer concerned with making use of the tensile strength of cast iron will most probably have been aware of a feeling of uncertainty, particularly when extensive use had to be made of this property.

Nowadays, however, there are available to the designer grades of cast iron having a tensile strength of $35\,kg/mm^2$ ($22\,tons/in^2$), and even a great deal higher. This means that cast iron can now offer the designer a material which in respect of tensile strength approaches very closely to ordinary structural steel. Therefore, all that is needed in so far as lightweight construction is concerned is to design in such a way as to utilize these properties to the full.

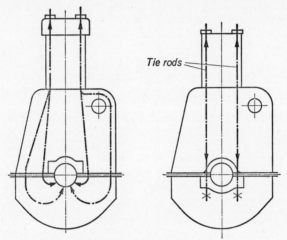

Fig. 245.—Satisfactory disposition of lines of force
through tie rod construction

Lightweight construction in cast iron is governed by practically the same factors as for construction in steel. In the first place the designer must try to eliminate all notch effects as far as possible. All transitions must therefore be carefully designed, that is to say the change of shape must be made gently (see pages 81–2) not only for purely casting reasons but also out of strength considerations.

By adopting a suitable general arrangement or layout the designer always has it in his power to make certain that the lines of force do not take on a complicated pattern but run in paths of minimum curvature.

The designer allows for the difference of behaviour under tensile and compressive loading by using the unsymmetrical cross-sectional shapes ⌐, ⊥, Λ. It can then be reckoned that the risk of fracture will be about equal on either side.

The hollow form should be used whenever torsional loading arises.

The tubular form is naturally the most satisfactory, as shown previously in fig. 238.

If at all possible, therefore, the designer should use closed sections exclusively and should avoid ribs. The closed type of construction is significantly stiffer and affords considerable weight saving. For the purposes of lightweight construction therefore the types illustrated in fig. 246 are recommended.

It may happen of course that a closed construction cannot be used. If this is so, recourse must be had to ribbed shapes. It should be noted, however, that the form of rib used will depend on the prevailing loading conditions. Deep ribs certainly confer maximum load-carrying capacity,

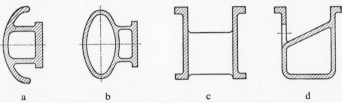

Fig. 246.—Cross-sectional forms for lightweight construction in cast iron

(*a*) and (*c*) unsatisfactory   (*b*) and (*d*) satisfactory

combined with minimum deflection when static loading exists, but their ability to absorb impact energy is very small. There is also a risk that stress peaks will be set up. A plate devoid of ribs has a very large capacity for absorbing impact loads. This means that shallow and fairly wide ribs should be preferred (fig. 90) in any situation calling for maximum load-carrying capacity combined with the ability to absorb fairly large amounts of deformation.

If all these factors are taken into account, it is perfectly possible to achieve in castings also such a high standard of form-derived strength that they are able to enter into competition with designs in steel. One need only recall the cast-iron crankshaft designed on the above principles.

The designer with a problem in lightweight construction should therefore remember in the first place that the lightweight principle forms the basis for his calculations and for all the later stages of designing. The factors which he should bear in mind are as follows:

1. Try to clarify the static and dynamic force conditions as completely as possible.
2. Try to secure a maximum symmetry in force application.
3. Provide suitable supporting and mounting arrangements.
4. Choose a method of construction offering maximum compactness.
5. Design for the most favourable possible flow of forces.
6. Ascertain points giving rise to notch effects and eliminate them.
7. Use suitable sections.

8. Use hollow or cellular construction.
9. Stiffen flat surfaces by suitable supports, swages, and ribs.
10. Take into account all factors related to fabrication by welding.

Exercise problem

Problem 29

The frame of a riveting machine is usually made in cast steel and has the shape shown in fig. 247. The greatest stresses occur in the curved portion. Over this portion of the frame the strength calculations can only be carried out approximately for particular cross-sectional planes; they yield large dimensions.

Try to design the frame so that the load on it can be referred to one of the familiar configurations. This will make the calculation process simpler and clearer. All that is needed is suggestions for an economical design without going into actual calculation.

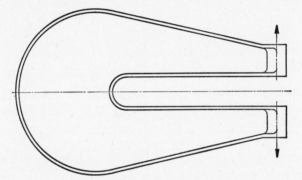

Fig. 247.—Frame of a riveting machine

(b) Best possible estimate of strength

The idea of using the materials as efficiently as possible, through the ability to predict the stresses accurately, is only feasible if the form of the members is very simple. In this case the real loading conditions can be made sufficiently near to the ideal for simple calculations to give a realistic result.

The kind of result which is wanted cannot be reached by intuition, nor merely by comparison with other cases; neither data drawn from experience nor empirical formulae will provide what is required. The thoughtful designer who uses such formulae will always be conscious of a sense of uncertainty for the very reason that the origin of the formulae is often not known to him. He never knows therefore to what extent the material is being utilized strength-wise, nor how the stresses are distributed over the cross-section, nor where the greatest stresses occur. Under such conditions a lightweight design, based on economical utilization of material, cannot be attained.

Classical strength of materials clearly is of great help to the designer.

It enables him to calculate the dimensions of components on the basis of a sound theory. But even the beginner knows that such calculations only apply for simple forms such as straight parallel or prismatic bars, bars bent in circular arcs, flat or dome-shaped discs, rings, or tubes. It is here, however, that difficulty arises, since it is only very rarely possible to derive the forms of structural members from these simple elements.

The designer is thereby obliged to make assumptions about stresses in parts which often approximate only very roughly to simple basic forms. It is not surprising therefore that such parts sometimes carry stresses exceeding those found by calculation, so that failure occurs.

There are further factors, however, which introduce uncertainty into the calculation. The elementary theory of the strength of materials tells us that tensile and compressive stresses are distributed uniformly over the cross-section and that bending and torsion stresses increase linearly from the neutral axis. There are factors which interfere with these distributions and which have led to the inclusion of a safety factor in the calculation. If one considers, however, that this factor may range up to a figure of 20, especially when dynamic loading is involved, and that it often varies by as much as 50 per cent for a given loading condition, it will be seen that no one using such factors could possibly claim to be employing material-saving or " lightweight " methods of construction.

As mentioned in the previous section, the occurrence of recesses, key-ways, bores, and any abrupt changes of cross-section, sets up stress peaks or notch effects which are considerably more severe than the nominal stresses distributed uniformly over the cross-section. In recognition of this fact, and on the basis of many experiments, strength theory has been developed to allow more fully for shape effects. The most important fact is that strength depends not only on the material but also on form, type of loading, and time. It is important that the student designer should familiarize himself with modern strength theory, because it provides the only key to calculating machine components more accurately, that is to say in accordance with weight-saving principles.

The disposition of the lines of stress shows the designer where the greatest stresses (stress peaks) occur. Unfortunately it is rarely possible to draw the stress pattern for a newly developed machine component on the basis of simple rules. Instead, the designer is obliged to study existing known designs so that he may gain a feeling for the probable stress pattern.

The magnitude of the stress peak is expressed numerically by the form factor, which indicates the ratio of the greatest stress to the uniformly distributed nominal stress. These form factors have already been determined by experiment for a considerable range of simple constructional elements and are available to the designer in the literature of the subject.

In the absence of any guide to the true distribution of stresses in a workpiece, the only course left is to make direct measurements.

In all other respects the calculation procedure is subject to the same considerations as discussed previously in the section dealing with the influence of mechanical stressing on form design (see page 72). The principal factors may be summarized as follows:

1. Determine the optimum form in the first place by using only the laws of the classical theory of the strength of materials.
2. Reduce stress peaks (see page 194) after determining points giving rise to notch effects.
3. Carry out the final calculation and dimensioning in accordance with the theories of fatigue strength and strength as a function of form.

(c) Use of welding instead of riveting

During the discussion on the form design of welded fabrications (p. 127) it was seen that welded joints yield a much more favourable stress

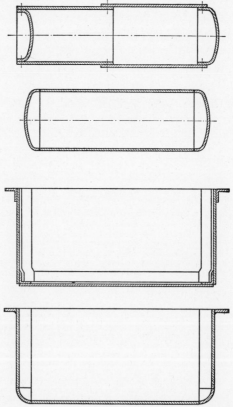

Fig. 248.—Comparison of riveted and welded fabrication

pattern than riveted fastenings. This makes it possible to dispense with all the aids used in riveting, such as straps, laps, stiffening brackets, and (possibly) gusset plates. This feature means reduced weight and argues for the use of welding wherever a lightweight mode of construction is required.

The advantage gained can be seen most easily from the examples given (fig. 248).

The straightforward stress pattern obtained with welding means that the strength conditions are much more favourable and also more amenable to calculation. There is no need whatever to design sharp edges

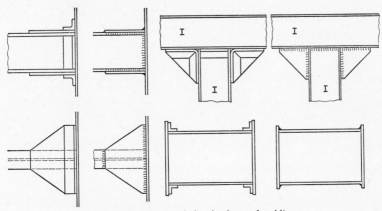

Fig. 249.—Joints made by riveting and welding

into flat-surfaced constructions such as bins and tanks which are to be fabricated by welding. Such edges are always sites of exceptionally severe loading. In contrast with riveted construction the corners can be lightly rounded, with the result that the structure benefits both in strength and stiffness.

In building with lattice work, the principle of lightweight construction has always been applied. Clearly defined force conditions are created (through the medium of a force diagram) and only tensile and compressive forces are used. The structural material is then positioned along the lines of these forces. Experience has shown that even here it is still possible to save weight through welding. This is true to a much greater extent for lattice structures made of tube, since with this type of construction there is no scope for riveting.

Welding affords a greater saving of weight, however, when $\mathbf{I}$-beams or plate construction are used. The stiffening brackets become redundant, and designs which are both light and elegant result (fig. 249).

The principal advantages of welding over riveting can therefore be summed up briefly as follows:

1. Low weight   4. Greater strength
2. Low-cost manufacture   5. Greater stiffness
3. Smooth surface (plain appearance)

Exercise problem

Problem 30

Re-design the riveted frame shown in fig. 250 for fabrication by welding in such a way that an appreciable saving of weight is obtained.

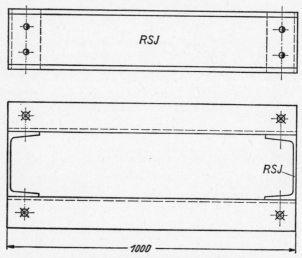

Fig. 250.—Riveted frame. Weight 78 kg

(d) Use of welding instead of casting (lightweight construction in steel)

For the designer there can be no such thing as preconceived opinions. Every step he takes must be dictated by economic factors. This applies equally to the choice of a suitable material. It must be the aim in light-weight construction to reduce weight per unit of performance.

The factor which determines all the designer's efforts is therefore the ratio

$$\frac{\text{applied resources}}{\text{performance}}$$

where applied resources covers not only weight, but also the volume of the material, the amount of power required, maintenance cost, manufacturing cost, transportation cost, etc. In some instances weight and amount of space occupied play the most important part, as in automobile and aircraft engineering, whereas with machine tools and agricultural machinery, for example, all the other factors under the heading of applied resources assume more or less equal significance.

Welding, and electric welding in particular, has made it possible for the designer to substitute steel fabrications for castings on a very extensive scale indeed. On the other hand, it is necessary to avoid falling into the error of thinking that the use of cast iron must be avoided in all circumstances. We have already seen that it is readily possible to design for lightness in cast iron (see page 198). Furthermore, casting has the big advantage of offering almost unlimited freedom in design so that, for example, small and very intricate engine casings involving many features and details can be made more cheaply and lightly in cast aluminium than in steel.

It cannot be emphasized too often that the designer must do justice to the material and the manufacturing method concerned; in other words, the form of a component made of steel will be based on principles differing from those governing the use of cast iron. The tensile and bending strength figures of ordinary cast iron are scarcely half as high as those of ordinary structural steel, and this fact alone allows the designer to keep dimensions smaller when designing in steel. The modulus of elasticity of cast iron, too, averages less than half the figure for steel, so that, assuming the same distance between supports, the same loading, and the same moment of inertia, the deflection obtained with cast iron will be about twice as large as that of steel. Assuming identical conditions, therefore, castings do not yield greater stiffness than steel members. By using steel it is possible to make do with smaller wall thicknesses (about half the thickness required for cast iron), whilst at the same time the structure benefits from increased stiffness.

When one considers that castings are usually over-dimensioned, not only to impart strength but also for reasons of casting technique, it is easy to understand that construction in steel is capable of giving a weight saving of 50 per cent and more compared with casting. Added to this, however, is the fact that welding allows all lugs and other reinforcing and stiffening features to be dispensed with.

Construction in steel therefore offers a whole range of advantages over construction in cast iron.

1. Low weight.
2. No risk of shrinkage-cavity stresses and casting stresses (homogeneous material).
3. Greater strength and stiffness.
4. Capacity for subsequent modification.
5. Greater denseness, even with very small wall thicknesses.
6. Smaller machining allowances or complete elimination of such allowances.

7. Price reduction brought about as follows:

  (*a*) Saving on patterns and all associated costs including storage, maintenance, dispatch.

  (*b*) Saving on freightage owing to lighter weight.

  (*c*) Shorter delivery time.

  (*d*) Avoidance of rejects.

  (*e*) Saving on machining.

  (*f*) Small capital investment for workshop equipment.

Exercise problem

Problem 31

Re-design the bracket, originally made in cast iron, for fabrication by welding in such a way as to give maximum saving of weight. Observe the dimensions indicated in fig. 251.

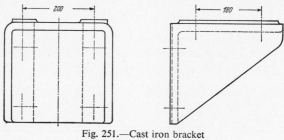

Fig. 251.—Cast iron bracket

Weight approx. 28 kg

It often happens that existing components in the form of castings have to be re-designed for manufacture in steel. The best way to set about this

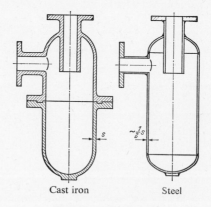

Cast iron   Steel

Fig. 252.—Oil separator

is to reduce to half their original values all wall thicknesses subject to tensile and bending loads (fig. 252). Apart from the reduced wall thicknesses obtained in this way, there is a further saving of weight, in the example illustrated, owing to the fact that the flange needed with the sub-divided type of construction are eliminated. Similar savings of weight are obtained when a cast iron gearcase is re-designed for fabrication in steel (fig. 253).

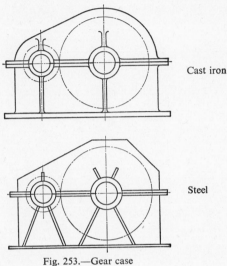

Cast iron

Steel

Fig. 253.—Gear case

In the above examples the shape used for the steel version was closely modelled on the shape of the original casting; on the other hand, if welding reasons make it necessary to give the steel version a form independent

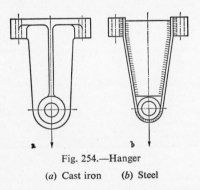

Fig. 254.—Hanger

(*a*) Cast iron   (*b*) Steel

of the cast version, then the rule of thumb which recommends halving the wall thickness for fabrication in steel can no longer be applied. In this situation (fig. 254) the calculation must be carried out in a manner suited to the conditions resulting from the new shape. In the example shown in fig. 254*b* the strap takes the purely tensile forces. The plate inside is provided only for stiffening purposes.

The form design of steel parts is subject to all the factors discussed earlier when the influence of form on lightweight construction was discussed. These considerations apply particularly when designs resistant to bending and torsion are required.

(e) Use of high-grade steel

Every beginner knows that an increase in the permissible stress on a material results in smaller dimensions. Therefore, if, as is true of steel, the specific gravity remains virtually constant for all grades, the result of using a high-strength steel will be a decrease in weight. The use of steel of a higher grade would therefore be a simple method of achieving a lighter mode of construction. The difference could be quite considerable. It is only necessary to note that high-tensile bolts may carry four times the stress of mild-steel bolts and so give a saving of weight of 75 per cent.

The beginner may therefore well ask: Why not use the high-grade material right from the start for all purposes? However, the situation is not as simple as that. Of course the higher-grade material offers greater fatigue strength, but the strength data on their own are not all-important. We know that choice of material is influenced by a whole range of properties, such as hardness, elongation, toughness, notch sensitivity, impact strength, as well as by workability, manufacturing cost, and, naturally, by the price of the material.

The high-grade steels also have disadvantages which the designer cannot afford to accept without due consideration. First of all there is the fact that notch sensitivity becomes more pronounced with increasing strength. It is therefore only possible to make use of the improved fatigue-strength values provided that the surface is given a suitably high-grade finish after the stress peaks have been satisfactorily reduced. In addition, the high-grade steels give smaller elongation figures than ordinary machinery steel. Whether or not the designer can dispense with these properties will be decided by the requirements he has to fulfil.

The capability of being worked is a further factor which must be considered. High-grade steels are often either unweldable or weldable only to a very restricted extent. Therefore if the designer decides to use a steel which is difficult to weld he must be prepared to do without the design opportunities afforded by welding.

There is a further point which needs to be emphasized. When a high-grade material is used, it may very easily happen that the gauge becomes light enough for there to be a tendency for flat surfaces to buckle or bulge.

On no account should design be pushed to such limits through the use of high-grade materials. Of course remedies can be provided in the form of ribbing, swaging, and other methods of stiffening. In situations of this kind, however, the designer would be better advised to resort to light metals, because by so doing he will obtain structures which are both larger and stiffer for the same weight. It will be seen therefore that there are limits to what can be achieved with steel. More will be said on this subject in the next chapter.

Before he decides to use a higher-grade steel, the designer must consider very carefully whether he will in fact gain from doing so. It may be that the advantage gained will not justify the cost. The designer who takes this course will find that the only answer is to obtain comparative figures.

Exercise problem

Problem 32

A gear made of cast steel having 60 teeth, a module of 6 mm and facewidth of 60 mm is to be scaled down in size and weight by the use of a higher-grade material whilst preserving the same number of teeth and the same width ratio. Assume a circumferential force of 550 kg at 150 r.p.m.

(f) Use of lightweight materials

To the uninitiated the term " lightweight construction " first suggests the use of structural materials lighter than either cast iron or steel, such as the aluminium alloys or plastics. Anyone who has been concerned with lightweight construction, however, knows that more weight can be saved by " lightweight form design " than by using a lightweight material.

A profitable exercise for the beginner is to make comparative calculations with a variety of constructional materials so that he may get an idea of what is attainable with each. Fig. 255 gives a comparison of various materials with steel on the assumption of a circular cross-section and identical loading conditions.

From fig. 255 it can be seen that the specifically lighter material, e.g. wood, does not necessarily afford any appreciable saving of weight, quite apart from the undesirable characteristics and the relatively very large dimensions in comparison with steel. We have already learned that a cast-iron component involves more weight than its equivalent in steel. If, however, a steel offering higher strength had been used for comparison purposes, then the size in Duralumin would have been larger.

The criterion for the reduction in weight is the

$$" \text{ specific strength by weight } " = \frac{\text{permissible stress}}{\text{specific gravity}} = \frac{\sigma_B}{\gamma}$$

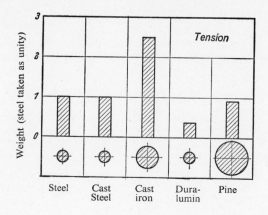

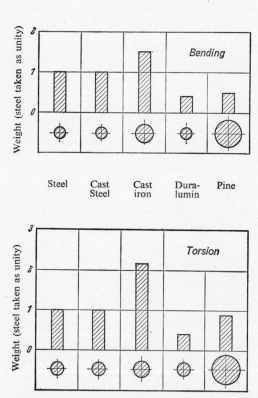

Fig. 255.—Weights of various materials needed to sustain the
same load as steel

Fig. 256 very clearly brings out the advantage of using light metals instead of steel. If a high-grade aluminium alloy (point 1) and a given steel grade (point 2) are selected along the straight lines representing the materials in question, then the saving of weight is given by the ratio

$$\left(\frac{\sigma_B}{\gamma}\right)_{St} : \left(\frac{\sigma_B}{\gamma}\right)_{Al}$$

When the steel has the same tensile strength as the aluminium alloy (points 3 and 1) the weight saving obtained is given by the ratio of the specific gravities. As the tensile figures of the steel go higher there will continue to be a saving of weight on the aluminium side to an extent given by the ratio of the specific strengths by weight. Not until a very high-grade alloy steel is reached (point 4) that is to say when

$$\left(\frac{\sigma_B}{\gamma}\right)_{St} = \left(\frac{\sigma_B}{\gamma}\right)_{Al}$$

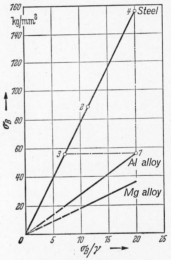

Fig. 256.—Comparison of specific strengths by weight

does the weight ratio become unity so that steel and light metal both give the same structure weights.

If the answer to the steel versus light metal problem depended on nothing but the tensile-strength figures, it would be possible to define the boundaries of the applications of these two types of material on this basis. We know, however, that there are other properties which also determine the choice of material.

One has only to examine the tensile-strength figures to discover factors which limit the use of high-grade steels. The full utilization of high-strength steels usually gives rise to very thin-walled components, so that the necessary stiffness cannot be guaranteed.

The use of light metals necessarily entails larger dimensions if the tensile strength is lower than that of steel; this means that greater stiffness and a safeguard against buckling and bulging are obtained right from the start. This is highly important in automotive body-building, because the thickness of light-metal sheet has only to be increased by a factor of about 1·4 to give the same stiffness and load-carrying capacity as steel.

The advantages offered by light metals under static loading are acknowledged. But nevertheless the province truly dominated by steel is the one occupied by stationary structures of considerable size, such as the frames or housings of machine tools, or the crankcases of heavy internal-combustion engines, the reason for this being that these types of structure do not allow full advantage to be taken of the benefits arising from the saving of weight.

The situation is different when movement and acceleration are experienced. The use of light metals under these conditions yields economies which derive from the reduced weight. A fan impeller may be taken as an example of this. When built in light metal instead of steel, the weight of the blades, and consequently the centrifugal force, is reduced in the same ratio as the specific gravities (to about one-third). From the outset, therefore, the conditions favour the use of light metals rather than steel. Similar advantages arise regarding the use of light-metal pistons and connecting rods.

Self-supporting truck bodies built of light metal can offer a saving of weight of as much as 50 per cent compared with the steel type. Consequently, the payload can be increased by the amount of this weight saving, or the payload can be kept the same and the benefit obtained in the form of reduced engine power, with resulting economies in fuel, tyres, etc.

A practical example of the benefits conferred by light-metal construction has been described by O. Suhr. An overhead travelling crane of 10 tons capacity and 23·16m span was built in steel and in an Al-Cu-Mg alloy of 40kg/mm^2 (25tons/in^2) tensile strength. Although offering the same degree of safety and stiffness, the light-metal version weighed only 19·5 tons compared with 36·3 tons for the steel version. The saving of material weight was therefore 45 per cent. This in turn, however, led to further advantages of a kind which the designer should not fail to take into account.

The weight reduction had the effect of reducing the power demand by 50 per cent. In addition, it was possible for the same reason to increase the travelling speed by 30 per cent. This, however, does not exhaust the benefits obtained. Owing to the reduced weight the craneway bearers can be made lighter. Furthermore, the stanchions supporting the craneways can be designed on less substantial lines. A further advantage, and one which should not be underestimated is the elimination of protective painting. It is easy to see that the initial cost, even if higher than that for steel, will be paid off more quickly.

Before switching to construction in light metals, therefore, the responsible designer must again consider which material will guarantee the best overall efficiency measured as

$$\frac{\text{applied resources}}{\text{performance}}$$

The lower fatigue strength and greater notch sensitivity of light metals might well give rise to hesitation in so far as their use for dynamically loaded parts is concerned. Experience has shown, however, that even under conditions of alternating stress the use of light metals still offers advantages on a weight basis.

There is one point which deserves special emphasis in this context, namely that the designer engaged on making the calculations for aluminium parts cannot simply proceed in the same manner as for steel. Owing to the lower modulus of elasticity the deformations which occur in light metals are considerably larger. This means that when designing in light metals it is necessary to check deformations at points which would not be checked if the design were in steel. Thermal expansion is about twice as large for aluminium as for steel, and this factor must of course be given special attention when composite methods of construction are used. There is also a greater danger of elastic instability so that Euler values and slenderness ratios must be watched. Anyone concerned with designing in light metals would be well advised to acquaint himself first of all with the experience of others in calculating light-metal structures.

There are still other lightweight materials, such as magnesium alloys, plastics, and wood, which the designer can often use to advantage. The low specific gravity (1·8) of the magnesium alloys, combined with their other properties, makes them suitable for such automotive applications as oil-pump bodies, gearbox top covers, crankcases, and brake shoes. The mechanical and technological properties of the plastics and of wood show some marked departures from those of the light metals. When proper allowance is made for their properties, these materials can be used not only for lightly stressed parts in light engineering but also for heavier-duty parts such as gears.

(g) Use of special sections

The use of standard sections affords little advantage in lightweight construction. Quite apart from the low strength values (which are not even guaranteed) the cross-sectional forms of these sections have sharp corners which are unsatisfactory from the point of view of stress formation. Much more suitable are the lightweight steel sections with uniform wall thickness and radiused corners. Sections of this kind are available to the designer in profusion. By making use of these sections it is possible to build automotive chassis members which are lighter in weight than corresponding parts built with standard sections. An example has been quoted of a longitudinal member for an agricultural trailer (fig. 257). The saving of weight obtained by building this member on the lightweight

principle with special steel sections was 34 per cent compared with the
design using standard sections.

The extrusion of light metals permits much greater diversity of cross-
sectional form. These sections are used by the designer with outstanding

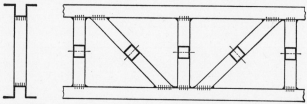

Fig. 257.—Longitudinal member built on the lightweight principle

success for frame and other applications in the automotive field. The
comparison of two types of pillar cross-section and window rail shows the
big advantage obtained by using special sections. The complicated built-up
type of construction involving many separate parts, expensive manufacture,
and costly assembly work can be greatly simplified and lightened by the use
of special sections designed for the purpose (fig. 258*b*).

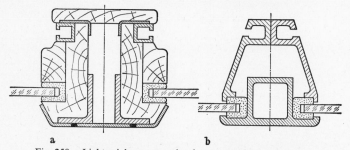

a   b

Fig. 258.—Lightweight construction by means of special sections

(*a*) Steel version   12 parts   Weight: 11 kg/m
(*b*) Light-metal version   3 parts   Weight: 3·6 kg/m

(h) Use of new components

There is a whole range of components which enable space and material
to be saved by suitable design methods. The designer makes use of such
components for purely economic reasons, even when he is not primarily concerned with lightweight construction for itself. For the purposes of lightweight construction, however, these components are indispensable aids.

One of these space- and weight-saving components is the Seeger retaining ring. Apart from lowering material cost, it also has the advantage of
saving man-hours on assembly, etc.

The Seeger retaining ring (fig. 259) is a type of snap ring used for preventing endwise displacement of machine components such as pins, shafts,

washers, gears, and anti-friction bearings. These rings are thus able to take the place of other types of locating feature such as split pins, shoulders, washers and nuts, or collars. A publication available from the makers gives details of a number of typical applications. The amount of metal used is 15 to 60 per cent compared with layouts which do not make use of retaining rings. The important savings of weight and space made

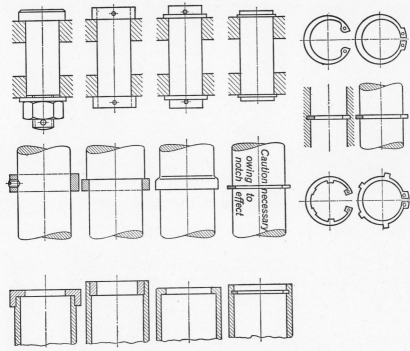

Fig. 259.—Use of Seeger retaining rings to save weight

possible by the use of Seeger retaining rings are illustrated in fig. 259. A further highly desirable feature is the marked simplification of design.

The familiar types of sealing arrangement for anti-friction bearings include felt washers, lip-type seals and labyrinth-type seals with or without thrower. These methods have the disadvantage of taking up a fairly large amount of space and often need more room than the bearing itself.

Now available, however, is a sealing device known as a *Nilos ring* consisting of a springy disc some few tenths of a millimetre in thickness. This device takes up hardly any room, and its weight is so small compared with other types of seal that one is almost tempted to talk of a 100 per cent weight saving.

Nilos rings are available in two types (fig. 260). It is only necessary for them to be held fast at *a* and for the edge *b* to bear against the bearing ring under a light pre-load. When correctly fitted these devices give a perfectly

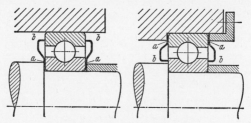

Fig. 260.—Types of Nilos ring

efficient seal. They are not suitable for use with self-aligning bearings and parallel roller bearings with endwise movement. If likely to be damaged, the rings must be protected by cover discs. Fig. 261 shows how the Nilos rings save space and weight.

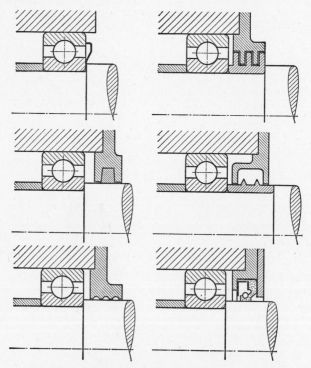

Fig. 261.—Comparison of Nilos ring with other types of
sealing arrangement

Important weight savings can be achieved with bolts by using high-grade material.

If bolts of mild steel are compared with high-tensile bolts it will be noted that, as indicated in fig. 262, the diameter of the high-grade bolt can

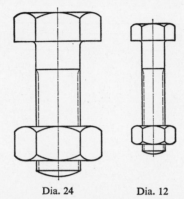

Dia. 24   Dia. 12

Fig. 262.—Weight saving obtained by using higher-grade steel

be reduced by half, and this in turn means that the weight is about 75 per cent less. The reduction of the bolt diameter also makes it possible to reduce all companion dimensions such as flange thicknesses and diameters so that a substantial saving of weight is obtained (fig. 263).

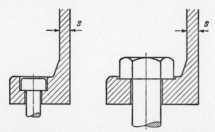

Fig. 263.—Space- and weight-saving design using high-grade bolts

Socket-head screws offer the designer great latitude. They permit flush screwed fastenings of pleasing appearance and are therefore specially favoured by machine tool builders. For lightweight construction, all bolts must be of high-tensile material.

The wiretrack ball bearing is an anti-friction bearing of exceptionally lightweight design which is space-saving and economical in its use of metal. It can be used for rotary and rectilinear motions, but is suitable only for relatively light loading and low rotatory or translational speeds. The ball

8

tracks consist of wires made of high-grade spring steel. The bearings can be adapted to any specific application by fitting the rings into specially designed housings.

The firm's publication should be consulted for details of calculation, design and assembly principles.

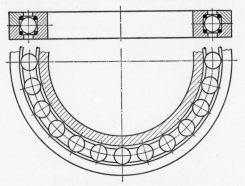

Fig. 264.—Wiretrack ball bearing according to Franke

(i) Saving of weight through basic change of layout

The choice of a lightweight form is the key to all further design steps for achieving the lightest possible weight. To some extent, however, weight depends on the choice of basic layout, and a change to an alternative design can sometimes yield opportunities of reducing weight still further.

Two examples given by F. Mayr in the *VDI-Sonderheft* 1942, page 27,

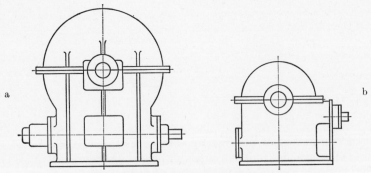

Fig. 265.—Weight saving achieved by changing the working principle
(according to Mayr M.A.N.)

(*a*) Single-stage reducer with 50-tooth worm wheel   Total weight: 843 kg
Efficiency: 65%
(*b*) Two-stage reducer with globoid tooth system
Total weight: 370 kg   Efficiency: 92%

may illustrate this. A single-stage worm gear reducer (fig. 265) with fifty teeth and a small lead angle had an efficiency of 65 per cent. A large housing was needed to dissipate the heat. The addition of an oil cooling-system would probably have enabled the housing to be made smaller with consequent saving of metal. Instead, however, this layout was abandoned in favour of a two-stage unit. A spur-gear reduction stage was provided in front of the worm gearing, and the globoid tooth system was adopted. In this way the weight was cut from 843 kg to 370 kg and at the same time the efficiency was increased to 92 per cent.

A further example is shown in fig. 266. When an electric motor is coupled to a driven machine or, as in this example, to a gear unit, then

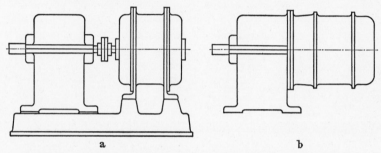

a   b

Fig. 266.—Lightweight construction achieved through a different layout

the equal and opposite torques must be reacted by a bedplate common to both units (layout *a*). If, however, a flange-mounting motor is used and is close-coupled to the gearcase, then the torques set up no longer act externally but are balanced within the unit. A bedplate is therefore no longer needed (layout *b*). This arrangement gives a marked reduction in the total weight.

In both examples a substantial saving of space is obtained in addition to the weight saving. That is why consideration of this aspect is recommended as promising success whenever a design is subject to the restraint of space restrictions.

(k) Lightweight construction by appropriate choice of parameter

A further measure contributing to space- and weight-saving design has already been mentioned on page 183 under point 4. It consists in making the best possible choice of the parameters which are involved. In this way the designer is able to make quite considerable savings of space and therefore of weight also.

A familiar example will illustrate this. The cylinder capacity V and therefore the size of a single-acting single-cylinder steam engine is given by the relationship

$$V \propto \frac{B.H.P.}{pn}$$

For a given rating therefore the size, and consequently the weight, of the engine is more or less inversely proportional to the indicated pressure p and the speed n. The designer thus has a ready means of keeping the size and weight of the engine small. He need only increase the initial pressure or the speed, either one or the other, or both together. These measures also have the effect of increasing the power to weight ratio.

It will be seen from this that the designer interested in lightweight construction has a variety of methods to choose from. He does not have to proceed by trial-and-error methods selected at random; instead, by using the methods discussed here he can work to good purpose towards a solution.

8. HOW THE USE OF STANDARD COMPONENTS INFLUENCES FORM DESIGN

It is scarcely necessary to enlarge on the advantages of standardization. Every designer accustomed to working on original tasks will be glad not to be burdened with having to solve each time afresh a whole range of constantly recurring identical problems. This explains why works standards of the most diverse kinds were in use long before a unified system of standardization was introduced.

At the time when work started on standards there were many designers who feared restriction of their freedom for original design. Experience has shown, however, that these fears were quite groundless. On the contrary, it was standardization which allowed the designer to devote himself to original work. Instances where expediency has called for a departure from standard specification practice are comparatively rare.

Standardization is always obligatory for the designer. Even during his training therefore it is important that his attention should be drawn to the various groups of standards which are distinguished as follows:

1. Basic standards dealing with abstract concepts, nomenclature, number sequences, screw threads, and fits.
2. Standards dealing with engineering drawing practice.
3. Material standards.
4. Standard components: bolts, nuts, pins, operating elements, etc.
5. Standards dealing with power transmission components, bearings, and gear tooth systems.

Limits and fits.—The designer learns correct dimensioning practice and the method of indicating tolerances in working drawings right at the start of his training. He is thus made aware, at a very early stage, that correct dimensioning and tolerancing govern accuracy of manufacture, the proper functioning of moving parts, and ease of assembly. The beginner experiences most difficulty in arriving at a correct choice of fit, that is to say in deciding the dimensional relationship of two parts which are required to fit together. From the bearing manufacturers the designer can obtain information on a very wide range of applications involving the fitting of rolling bearings into housings and on to shafts. Although this information serves as a guide, the designer must nevertheless consider when choosing the class of fit that other factors need to be taken into account, namely the loading conditions, facilities for fitting and removing, whether to use divided or non-divided bearing housings, whether removable or fixed bearings are required, and whether thermal expansion is likely to occur.

In any industrial firm the designer will have only a limited range of fits to choose from, because it would be highly uneconomical to hold an unduly large stock of standard gauges. The only advice which can be offered to the designer who is in doubt about the class of fit to use is to select the one which is just close enough for the application concerned. When the design is an entirely new one, no final decision on the question of fits should be taken until the matter has been discussed with the production department; this applies particularly when batch or mass-production runs are concerned.

Type standardization.—It was soon recognized that it would be very uneconomical to meet every wish expressed by the customer with regard to commercial engineering products such as pulleys or couplings. Any attempt to do so would greatly overload the drawing office and the workshops. To appreciate this it is only necessary to consider the many patterns which would be needed and which would probably not differ from one another in any important respect.

When planning a range of sizes of a given product it is important not to make the steps too small. Experience shows that a gradation of this kind is best suited to human needs when it follows a geometric progression. The designer engaged on work of this kind would be well advised to adopt standard numbers.

If, for example, the problem is to standardize a worm gear reducer, then the sizes of the units can be defined by the centre distance graduated according to the preferred number series of B.S. 1638, and each of the resulting types can be made in a range of ratios also graduated according to the same series—for instance according to the following example:

100	125	160	200	Centre distance
10, 12·6, 16, 20	10, 12·6, 16, 20	10, 12·6, 16, 20	10, 12·6, 16, 20	Ratio

For storage purposes, too, it is more economical to limit stocks to selected ranges of bolts, rivets, pins, etc.

9. HOW EXISTING PRODUCTS INFLUENCE FORM DESIGN

Efforts to rationalize production have led to the manufacture by specialist firms of a whole range of components and units which at one time had to be made by the firm that needed them. Every designer must therefore take account of bought-out items, even if they are only bolts, rivets, collars, pulleys, couplings, etc.

Every firm accumulates experience. It is therefore perfectly natural that the wise designer should keep to proven designs and should confine himself to progressing one step at a time in any development work. This means that the designer's work will often consist of adapting or converting proven components and assemblies to meet new requirements. Apart from offering a certain guarantee of success, this method is also advantageous from the economic viewpoint, since it enables savings to be made on pattern cost, tools, and fixtures.

Considerations of size limit freedom of design where fairly large components and assemblies are concerned, particularly if the designer has to aim at a specific form, such as a modern streamlined appearance. Even under these conditions, however, the designer will find it comparatively easy to reach his objective by only modifying the less important castings. Difficulty occurs only when a fundamental change is needed because the existing arrangement is incapable of further development. In this situation a new design is necessary, and this entails experimental and development work which is always a financial burden to the firm concerned.

10. HOW APPEARANCE INFLUENCES FORM DESIGN

Anyone who traces the development of an engineering product will note that its outward appearance often changes considerably over the years. These changes are brought about not only by technical developments, by progressive improvement of the product, by the simplifying of manufacturing methods, etc., but also in response to the aesthetic awareness underlying contemporary creative activity.

Doubtless designers have always endeavoured to make their work as pleasing as possible in appearance. It may be remarked, however, that today we smile at the appearance of engineering products over which our fathers and grandfathers enthused. One can still find the ancestor of the modern automobile—say, an old Benz car— aesthetically pleasing if one imagines it in the environment for which it was built. It is much the same with the earliest locomotives. When one studies a bearing bracket (fig. 267) dating from the seventies of the last century, however, one experiences an uneasy feeling. This arises from the fact that the form reminds one of a wooden grandfather clock. The form is therefore alien, since it is not in keeping with the properties of the material nor with the methods of working it.

The idea of form is a measure which our senses apply to solid objects. In the example just quoted, however, there is a deception. Form can only be beautiful and possess an inner strength if it corresponds to the truth. Saint Thomas Aquinas expressed this idea by saying that: " Beauty is the gleam of truth ".

Fig. 267.—Bearing bracket dating from 1870

Attempts have often been made to trace in the literature the laws underlying the beautiful and pleasing form of some engineering design, and this has led to recognition of the fact that *everything which fits its purpose is beautiful*. This definition is in accord with the designer's intention, which is to impart to his work the most appropriate form with regard to function, material, manufacturing method, etc.

Despite designers' efforts to observe this principle, the outcome of their work in the engineering field often falls rather short of the ideal expressed in the definition. How does this come about? When two machines of equal performance and quality are compared with each other—and exhibitions afford plenty of opportunity for doing so—it may be found that one of the machines satisfies our aesthetic sense more than the other. Once this has been established, one soon notices that the more pleasing machine is also the one which is better fitted to its purpose by reason of one or more features which may include, for example, a more convenient grouping of controls, a less cluttered shape giving a neat and clean impression, or more pleasing paintwork.

There may also be another reason. Form is determined not only by manufacturing methods but also by design calculations. Every beginner knows that dimensioning is based on ideal configurations entirely, and it is often left to the designer's feeling for form to discover the correct one. This feeling for harmony can often be developed to a certain pitch, but there must be some inborn talent for it. Very probably this is also the reason why one designer achieves greater harmony in his work than another.

An example of the fact that fitness for purpose does not in itself entirely satisfy our aesthetic sense is afforded by the old type of Bosch horn illustrated here (fig. 268). Although all four types are perfectly suited to their purpose there is no doubt that the version shown in fig. 268*b* is the most pleasing.

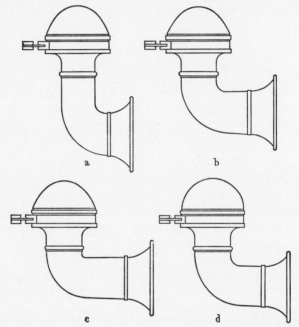

Fig. 268.—Various forms of the old Bosch horn (according to Kollmann)

Many firms make use of scaled-down wooden models in order to find the most pleasing shape for their products. In a big locomotive works which had become famous for the harmonious design of its various types of locomotive, it was the practice to build wooden models on a reduced scale to help to assess the aesthetic aspects of design. Motor-car builders also make use of models to work out the final styling details, using the models as well for aerodynamic experiments.

To achieve the most pleasing appearance possible the beginner should first develop his design in accordance with the rules of this book. When this has been done, and not before, he should compare his product with others which serve the same purpose but have a more pleasing form. In this way he too will begin to acquire the right feeling for harmony and beauty.

The layman often has a good appreciation of beauty without being

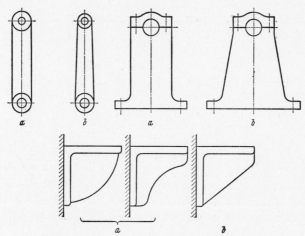

Fig. 269.—Satisfactory and unsatisfactory form design

able to explain the reason for it; and one can be sure that the designer has made a mistake if the layman fails to find beauty in the form produced. For example, anyone studying the forms shown in fig. 269 will find the types indicated at *b* more pleasing than the types indicated at *a*. The reason for this is that in *b*, in contrast with *a*, the strengths of the different cross-sections are about the same.

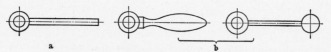

Fig. 270.—Convenient and inconvenient forms of lever

The lever forms shown in fig. 270*b* are more pleasing than the one at *a*. Why is this? Because the *b* types are better adapted to the hand.

Inaccuracies in manufacture always arouse a feeling of dissatisfaction; this is apparent from types *a* of fig. 271, whereas types *b* give an impression of accuracy and neatness and are therefore pleasing.

From the previous chapters we know that form is strongly influenced by the material and by the manufacturing method. It would be absurd to choose forms which are a contradiction of these requirements and of economic principles in particular. For example, the modern furniture industry uses nothing but plain smooth forms. This is because forms of

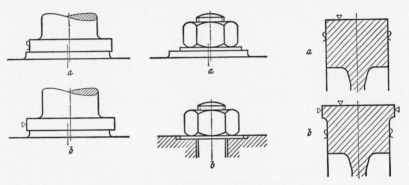

Fig. 271.—Designing to give an impression of neatness

this kind are easier and cheaper to make by machine methods than designs with carved features. Plain shapes are only justified in engineering products because they can be made at low cost and because they make for simplicity of maintenance and cleanliness in operation.

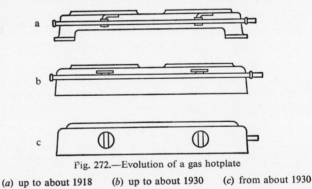

Fig. 272.—Evolution of a gas hotplate

(*a*) up to about 1918   (*b*) up to about 1930   (*c*) from about 1930

The present-day sense of form prefers plain smooth lines and even rejects over-emphasized radiusing. With regard to the appliances and devices used in everyday life this is very much to be welcomed. The gas hotplate illustrated here (fig. 272) shows how great are the changes which have taken place over the years by engineering products of this kind.

Hotplate *a*, typical of pre-1914 designs, was a ribbed casting with a black finish. It was therefore unattractive in appearance and difficult to clean. By the time of type *b* white enamelling had already been introduced, whilst the present-day model is perfectly plain, minus exposed pipe, white-enamelled, and provided with a cover. In its latest form the appliance gives a very neat and pleasing impression.

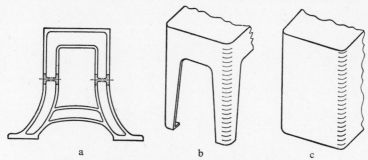

a   b   c

Fig. 273.—Present-day preference for plain smooth shapes

(*a*) Old style   (*b*) and (*c*) Modern style

Ribbed castings are always dirt traps. That is why the frames of modern machine tools are designed on plain smooth lines which are considerably more pleasing in appearance and no more costly to make, provided that

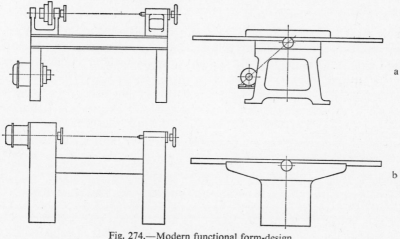

a

b

Fig. 274.—Modern functional form-design

(*a*) Old style   (*b*) Modern style

the right principles are observed (fig. 273). For very dusty operating conditions, as found in the woodworking industry for example, machines designed on plain straightforward lines offer special advantages (fig. 274).

Painted finishes play an important part. Greens and bluish greys have beneficial psychological effects. They tend to raise morale and make work more enjoyable. This accounts for the disappearance from the machine shops of the regulation black and dull, lifeless, uniform grey which were at one time favoured. Even structural steelwork nowadays is painted in pleasing colours.

The styles adopted by machine designers at different times indicate the changes in their ideas about appearance. Style in the purely aesthetic sense always demands uniformity of character. Any offence in this direction

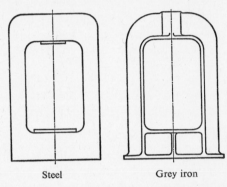

Steel   Grey iron

Fig. 275.—Simple pleasing shape of welded fabrication

upsets aesthetic sensibility. It is therefore not advisable to take an engineering component or structure of primarily cast construction and add parts by welding, for example by attaching welded or cast-aluminium brackets to a cast-iron housing.

The joining together of welded and riveted construction is offensive aesthetically. Welding differs from riveting in allowing the fabrication of much more elegant joints and supporting structures (fig. 249). Even pieces designed as castings can be made pleasing when fabricated entirely by welding (fig. 275).

11. HOW CONVENIENCE OF HANDLING INFLUENCES FORM DESIGN

The operating elements used on machines, appliances, and apparatus consist of handles, handwheels, star handles, and crank handles. These elements must be matched to the purpose they have to fulfil and to the magnitude of the forces or torques involved. All such parts, with the exception of heavy crank handles, are standardized.

Handwheels, and more particularly crank handles, are suitable for large forces. The torque required to operate such elements must be matched to the physical strength of the operator, and where necessary to his endurance also. Empirical figures are of course available on this subject. On the other hand, for making fine adjustments on machine tools, knurled nuts, star knobs, or small handwheels are used. Ball-type handles are widely used for limited rotary movements and for locking and clamping purposes.

To ensure safe and speedy operation the designer should arrange that the direction of motion matches the direction of the control initiating it. For example, the leftward and rightward slewing of a crane should take place in response to handwheel movements in the corresponding direction. For machine tools the principles governing the various motions are laid down in standards.

The safety of the operator must be looked after. Operating elements must be so arranged that no injury to the hands can result from using them. Ample clear space must be left around controls, and no projecting parts such as nuts, split pins, and sharp corners, permitted anywhere near them.

The operation of machines and mechanical equipment often exposes the operator to serious risk, particularly if he becomes less attentive or is distracted in some way. Safety regulations exist for this purpose, and it is the designer's business to know about them and apply them in his work.

12. HOW MAINTENANCE QUESTIONS INFLUENCE FORM DESIGN

The customer sometimes lays down certain requirements regarding the maintenance of a machine or piece of equipment. Even if he does not do so, however, the designer must take into account the factors mentioned previously in page 32.

13. HOW THE QUESTION OF REPAIR INFLUENCES FORM DESIGN

Provision for making repairs may influence form design quite considerably in certain circumstances. The points discussed in page 32 should therefore be considered carefully.

14. HOW SURFACE PROPERTIES INFLUENCE FORM DESIGN

The surface properties specified for engineering products vary with the type of duty concerned. For example, it may be necessary to provide resistance to chemical attack or a specific degree of hardness, or merely a certain type of appearance. The designer who wishes to use metallic coatings as a protection against corrosion should know that there are several such processes which, although primarily dependent on the type of material used, cannot be applied without regard to the shape of the work. This means that the designer must decide what type of metallizing process he is going to use and design accordingly. A similar situation exists with regard to hardening. The form of the work determines the type of hardening process employed and also influences distortion and fatigue strength.

15. HOW FITNESS FOR SHIPMENT INFLUENCES FORM DESIGN

The beginner tends to overlook this point. To save needless repetition reference should be made to the remarks in page 31.

16. HOW POWER REQUIREMENTS INFLUENCE FORM DESIGN

For purely financial reasons alone the customer will always be interested in keeping the energy demand low.

Manufacturer and customer have an equal interest in taking steps to keep the energy demand of a machine or piece of equipment as low as possible. As mentioned in page 32, the designer must ascertain the factors which have an important bearing on the reduction of energy demand, so that he can take them into account in his design work. Economy of consumption in the wider sense, including consumption of lubricants, cooling water, power absorbed in providing cooling air, etc., must also be aimed at.

G. COSTS

The only costs which the designer can directly influence by his work are material costs and labour costs. These, however, represent only a small part of the total cost. To them must be added of course the cost of purchasing, maintaining, and operating machine tool equipment, and the cost of running all workshop and office facilities and services. Other costs which must not be overlooked are the cost of office staffs, expenditure on social amenities and on advertising, etc.

All of these costs, apart from material and labour costs, are grouped under the heading of overheads, and they are considerably in excess of the two cost elements just mentioned. Overheads are also greatly influenced by the size and organization of the business and by the type of manufacture conducted, namely short-run or mass-production.

It is very difficult to allocate overheads to individual jobs with strict accuracy, since, for example, interest alone is a factor which is governed only by time. A common method of recovering overheads is to allocate to materials all costs associated therewith and to relate all other overheads to labour costs.

The total cost K can therefore be found from the following simple formula

$$K = BG_1 + LG_2$$

where B is material cost, L is labour cost, and G_1 and G_2 are factors representing overheads. Factors G_1 and more particularly G_2 may be of considerable magnitude (up to 800 per cent).

This formula should make it very obvious to the designer that total cost depends on keeping down material and labour costs, that is to say on correct economic thinking at the design stage. If the figure for total cost yielded by the final costing calculation does not agree with expectations, then the designer must seek the reasons and, as mentioned earlier in page 27, make the necessary changes in the form, manufacturing method, and material of the product, or even modify the basic design itself.

APPENDIX A

Solutions of exercise problems

Problem 1 (p. 36)

The designer engaged on solving a problem should not rely on inspiration to supply missing details. If he is accustomed to systematic working he will investigate the problem by studying the " customer's requirements " discussed in page 28.

1. *Functions to be fulfilled.*—The function is clearly indicated by the stated requirement which is to lift a certain quantity of water to a given height. The information regarding the speed of the pump is really superfluous; a customer would not be likely to specify this. The designer will determine the speed after weighing the economic factors. If the problem arises out of an inquiry, the designer will first check whether there is available a standard pump capable of doing the job. Only in exceptional circumstances will he be obliged to develop a new unit, which of course will entail considerable extra cost. If this is necessary, the customer will obviously have to be informed accordingly. If an entirely new design is found to be justified, however, there will be a whole series of questions to answer.

2. *Mechanical environmental conditions.*—Knowledge of environmental conditions is important to the designer. What is the cross-sectional size of the shaft? How is it lined? How can the pump unit be accommodated? Will there be sufficient space available for inspection and maintenance? These are questions to which the designer must have answers, otherwise he would be working in the dark.

3. *Mechanical loading.*—No information is needed regarding forces, since this will be obtained in the course of the design work.

4. *Climatic influences.*—If the unit is to be installed in a deep shaft it is to be expected that there will be considerable exposure to moisture; the designer will therefore have to take precautions against rust.

5. *Chemical influences.*—In view of point 4, it is unlikely that any other chemical influences will make themselves felt, since it can be taken for granted that only pure ground water is involved.

6. *Size and weight.*—Very restricted site conditions may impose severe limitations on the width of the unit. This is why further information on the dimensions of the shaft was called for in point 2 above. The weight factor is not in itself of importance but only arises in as much as it is considered in any rational design.

7. *Fitness for shipment.*—This factor is not likely to cause any trouble.

8. *Maintenance and operation.*—Detailed requirements should be notified in connection with these points so that the designer can express his views on the practicability of fulfilling them.

9. *Overhaul facilities.*—Agreement must also be reached on this point so that the customer will be spared disappointment over requirements which cannot be met.

10. *Economy of power consumption.*—It is a matter of the greatest concern to every customer that the equipment he orders should operate economically. It is the designer's responsibility to specify the driving power and the power demand of the installation as a whole. The customer may even ask for the figures to be guaranteed, but such guarantees can only be given on the basis of experiments.

11. *Service life.*—The question of length of useful life is often raised by the customer. If the firm has had long experience with similar units it will be in a position to commit itself on this point.

12. *Operating cost.*—This is closely tied to the question of economy of power consumption. The customer is therefore certain to raise this point.

13. *Freedom from noise.*—This factor is unlikely to be of any importance as far as a shaft pump is concerned.

14. *Appearance.*—This factor is unlikely to have any significance in the situation considered here.

15. *Delivery date.*—The question of delivery plays an important part in all orders. It depends for its fulfilment not only on proper planning of the drawing-office work, but also on economical working on the production side.

16. *Quantity required.*—The quantity required may have a decisive influence on design since, for example, casting would scarcely be entertained for a " one-off " or " two-off " order.

17. *Overall cost.*—This question is certain to be included in every inquiry. Its importance to the designer lies in the fact that it obliges him and his firm to plan as economically as possible owing to the competition they face.

From this example the student designer will appreciate the importance of checking an order by reference to the list of customer's requirements. In actual practice, too, this procedure will always be found worth while.

Problem 2 (p. 36)

We are told that this problem is in the nature of an inquiry. It is obvious that in practice there would be many points still to clarify in an instance of this kind. We will establish the essential requirements again by working through the familiar list of " customer's requirements " point by point.

1. *Action required.*—The problem concerns the lifting of a cover in such a manner that a basket can be lifted out. All necessary dimensions of the autoclave and of the cover are given. The designer will of course immediately ask himself the question: In what ways can the autoclave be opened? This is a very important point, because it may call for further information from the customer.

2. *Mechanical environmental conditions.*—These conditions are so clearly defined here that the cover can only swing up against the wall as there is no space for any

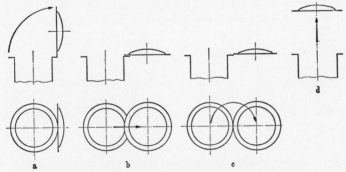

Fig. 276.—Possible solutions to problem of cover movement

alternative arrangement; in addition, it gives the best possible access to the autoclave (fig. 276*a*). The exact distance x from the wall (fig. 8) must be obtained from the customer.

Action
Mechanical environmental conditions
Mechanical loading
Climatic influences
Chemical influences
Size
Weight
Fitness for shipment
Maintenance and operation
Power consumption
Service life
Reliability
Operating cost
Operating instructions
Freedom from noise
Appearance
Delivery date
Quantity required
Overall cost

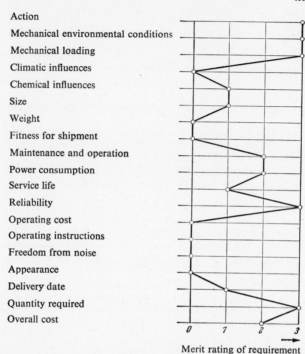

Merit rating of requirement

Fig. 277.—Presentation of priority ranking of customer's requirements

3. *Mechanical loading.*—The information in fig. 9 is sufficient to permit calculation of the cover weight which is the key factor in the dimensioning of the operating gear.

4. *Climatic influences.*—These are without significance, since the autoclaves are installed in covered premises.

5. *Chemical influences.*—It may be that the premises become filled with vapour from the soda solution when the autoclaves are opened. Protection against corrosion is therefore necessary.

6. *Size.*—Nothing is specified about this. The designer, however, will aim at the utmost simplicity and compactness.

7. *Weight.*—Unimportant.

8. *Fitness for shipment.*—This is not likely to present any difficulty.

9. *Maintenance and operation.*—The inquiry specifies manual operation. No further information is needed.

10. *Economical power consumption.*—Nothing is specified in this connection. The designer will naturally arrange the device so that its operation does not call for undue effort.

11. *Service life.*—No requirements are stated in this connection. Assuming operating conditions of ordinary severity the device should meet stringent requirements.

12. *Reliability.*—Although nothing is laid down about this factor no designer can ignore it without laying his firm open to a possible penalty. The application involved here makes it necessary that the cover should be held securely in all positions after it has started to lift.

13. *Operation cost.*—It can be said immediately that with manual operation this cost element will be very small if one disregards occasional lubrication of the bearings.

14. *Operating instructions.*—Operation is so simple that no instructions are needed.

15. *Freedom from noise.*—If any requirement had been expressed in this connection it would be possible to guarantee silent operation.

16. *Appearance.*—No special requirements to be met.

17. *Delivery date.*—The question of delivery arises in all orders and inquiries. In this instance the firm will fix delivery by arrangement with the drawing office and works.

18. *Quantity required.*—Four of the devices are required; there can therefore be no question of using castings.

19. *Overall cost.*—Obviously this is a question which arises every time. If there is no existing design to serve as a guide the costing calculation is usually based on the estimated total weight by applying empirical values.

A graph can now be drawn to show the priority ranking of the customer's requirements for the purpose of this particular problem (fig. 277).

Problem 3 (p. 49)

There are no supplementary requirements and the problem resolves itself into pure kinematics. A study of points 2 and 3 listed on page 38 soon leads to the solutions illustrated here (fig. 278).

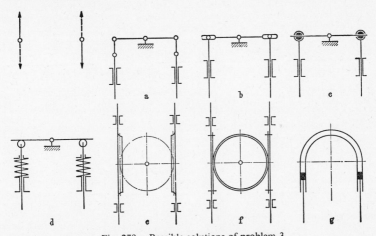

Fig. 278.—Possible solutions of problem 3

The problem to be solved is given at top left

Problem 4 (p. 49)

This problem is very simple and clearly defined. Despite this, however, the beginner is recommended to analyse the various requirements once again by reference to the list on page 28. In this way he will familiarize himself with the problem completely and will learn to identify the secondary requirements.

Analysed in the usual way the problem looks like this:

1. *Action required.*—Parallel raising and lowering of a table without rotary or sideways movement.

2. *Mechanical loading.*—100 kg load plus table weight.

3. *Climatic influences.*—No significance.

4. *Chemical influences.*—No significance.
5. *Mechanical environmental conditions.*—Indicated dimensions to be observed.
6. *Size.*—(fig. 21).
7. *Weight.*—No significance.
8. *Fitness for shipment.*—No significance.
9. *Operation.*—Handwheel with crank handle and horizontal shaft, fine adjustment.
10. *Maintenance.*—Nothing specified, no significance.
11. *Overhaul.*—No significance.
12. *Economical power consumption.*—Easy to operate.
13. *Service life.*—Nothing specified.
14. *Reliability.*—Table to stay in any position to which it is set.
15. *Operating cost.*—Nothing specified, no significance.
16. *Appearance.*—No significance.
17. *Delivery date.*—Not laid down by customer, therefore subject to arrangement.
18. *Quantity required.*—10.

In this problem so many of the conditions are already taken care of that it is possible to start by selecting from the various kinematic arrangements available. All that is needed in the first place is a list of possible solutions which take into account the prevailing conditions.

Plenty of suggestions for solutions (fig. 279) will be obtained by recollecting similar cases, such as arrangements for raising and lowering the tables of boring mills and milling machines or kinematic solutions to problems of like kind, and by systematically running through the various mechanisms.

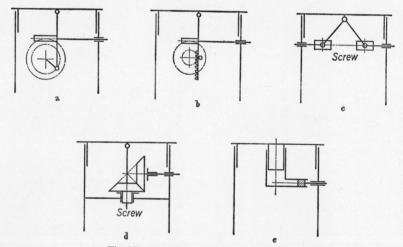

Fig. 279.—Possible solutions of problem 4

In the first place of course it will be the customer's requirements which will determine the solution selected as the optimum one. The designer, however, will add further requirements of his own, and these will be concerned with such points as economy of manufacture and stressing, which do not show up immediately in the problem as formulated, but which are nevertheless important in securing satisfactory operation and low-cost manufacture. For the purpose of evaluating the various solutions, therefore, we have available the customer's own requirements and the further requirements which arise from the need for an economical design.

We therefore have the following evaluation plan (fig. 280).

Solution	a	b	c	d	e	ideal
Fulfilment of function	2	3	2	3	3	3
Mechanical loading	3	3	3	3	3	3
Fine adjustment	0	3	0	3	1	3
Dimensions to be observed	–	3	–	3	3	3
Convenience of operation	–	3	–	3	3	3
Operating cost	–	3	–	3	3	3
Reliability	–	3	–	3	3	3
Manufacturing cost	–	2	–	3	2	3
Total	–	23	–	24	21	24

Fig. 280.—Evaluation plan

Solutions *a* and *c* can be disregarded because equal angular increments of handwheel motion do not give equal increments of table motion. As for solution *e*, the possibility of leakage losses occurring and upsetting the fine adjustment cannot be entirely ruled out. The only solutions left are *b* and *d*. The points ratings of these two solutions differ so slightly, however, that they can be regarded for the time being as equal in merit. On submitting solutions *b* and *d* to a further critical examination to assess straightforwardness of construction and probable cost of manufacture, it will be found that version *d* is the best.

Problem 5 (p. 49)

Solutions to this problem are to be found in books on kinematics. Anyone who takes the trouble to look through patent specifications will also discover a variety of solutions. As a matter of interest fig. 281 shows a number of solutions in the order in which they were patented. The year in which the patent was granted is indicated alongside each solution.

The first patented design for a " pendulum saw " to appear on the market is so delightfully simple that the outsider is surprised to discover that new designs featuring more complicated mechanisms were constantly being developed to succeed it. This may be partly because the firms in question wanted to have their designs patented.

Before an evaluation plan can be drawn up for selection purposes it is obviously necessary to find out more about the conditions involved, and about the customer's requirements in particular. Study of the various patent specifications yields the following requirements:

1. Ease of operation, i.e. easy forward and return travel.
2. Large cutting width.
3. Vibrationless running.
4. Maximum accuracy in straight-line movement of circular saw.
5. Lateral stability.
6. Restraining of any tendency for saw blade to climb.
7. Joints and guides to be protected against dirt.
8. Low overall height.
9. Simple construction.
10. Small amount of space occupied.

Once all the requirements have been properly defined—and this can only be done in consultation with the customer—the selection of the best solution can be undertaken.

One thing which is certain is that the patents shown here in chronological order do not represent a continuous general improvement on the original design. Instead, they all point to some specific feature in which progress has been achieved, for example: low overall height, easy-acting return motion, perfect straight-line movement.

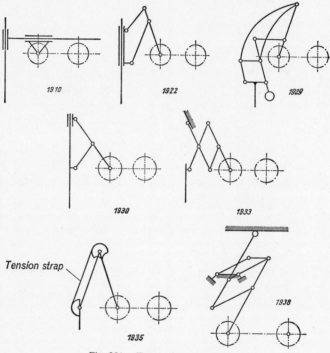

Fig. 281.—Types of pendulum saw

Problem 6 (p. 50)

When a beginner starts to design a worm gear unit and its housing from an existing model, it will probably not occur to him that the housing could be divided elsewhere, possibly with advantage. There can be no doubt, however, that he will benefit from considering all the possible methods of splitting the housing. Fig. 282 shows all the possible arrangements.

If the correct choice is to be made it will be necessary to state certain requirements which arise from considerations of satisfactory design and which have no direct connection with the customer's wishes.

These are the factors to be considered:

    1. Easy and accurate assembly without using dowels.
    2. No division through bearings.
    3. No division through oil reservoir.
    4. Ease of manufacture.

The best solution is easy to find even without using the evaluation plan. The best solution is in fact *f*. In contrast with *c*, it involves only lathework and makes for extreme accuracy in assembly. Solution *e* is less advantageous because here the dividing line runs through the worm bearings. Division on the lines shown in *d* would mean prior mounting of the gear in the case owing to the smallness of the bearing flanges. Types

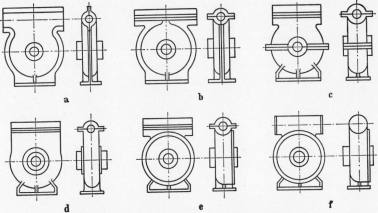

Fig. 282.—Various methods of dividing a gearcase

b and *c* are not easy to assemble. It will be seen therefore that the well-known example shown at *c* can be replaced with advantage by type *f*.

Problem 7 (p. 50)

In solving problems of like kind a systematic approach can be adopted, and all the factors affecting the customer's requirements and the operating requirements (page 28) investigated in sequence. On checking through these points, the only difference noted is that version *b* is more expensive than version *a* owing to the groove needed in the screw to prevent rotation. This means that, for the same diameter of screw, the stressing experienced by it is more severe in version *a* owing to the notch effect. Action, operation and energy consumption are the same for both types.

Problem 8 (p. 75)

First, find the external forces acting at joints B, C, D, E, F, and G (fig. 283). The given force *K* acts on the rod AEF at A. Of the other two forces *F* and *E*, the

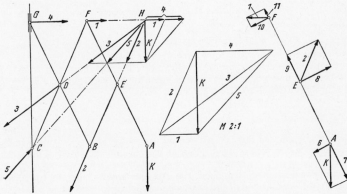

Fig. 283.—Resolution of forces

directions are known and consequently the magnitudes also (see force triangle) K, 1, 2. Force 2 at E can be displaced to B. In this way we obtain the external forces on rod GDB from triangle 234 and for rod CDF from triangle 135. The external forces acting on the various joints are therefore as follows: at E force 2, at F force 1, at D force 3, at B force 2, at C force 5, and at guide E force 4.

Apart from rod EB, all the other rods are stressed in bending, and tension or compression, as can readily be seen by resolving the external forces. For example, rod AEF is stressed in bending by forces 6, 8 and 10; whilst the EA portion of the rod is additionally stressed in tension by force 7 and the EF portion by force 11.

Problem 9 (p. 75)

The equation

$$\frac{Qa}{b} = R$$

gives the pressure R acting on the wheel and consequently the force which acts at point A of the arm (fig. 284). The external forces acting on the arm can then easily be found by the usual method to yield force G at B and force F at C. The latter acts as a tensile force on the spring.

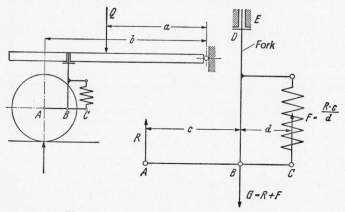

Fig. 284.—Forces acting on a single-wheel trailer

The external forces R, G, and F stress the arm in bending. The fork is stressed in compression over its entire length by force R. The lower portion is also additionally acted upon by compressive force G. Furthermore, the fork must sustain a bending moment Rc. The thrust bearing D is acted upon by the wheel pressure R, whilst the journal bearing is exposed to bending moment Rc.

Problem 10 (p. 91)

We know already that form design depends on many factors. In this example the only points to be considered are the forces involved and the material used, namely cast iron.

Since torsional loading is applied, the cross-sectional shapes to be considered primarily are the annular one or the box-section stiffened with diagonal ribs (fig. 285).

It would be a mistake to choose other types of cross-section such as the T-section or □, since these could only be given the dimensions necessary for adequate strength by using more metal. If a smaller reach were involved, then of course for simplicity's sake alone type *b* would be chosen, even without the diagonal ribbing.

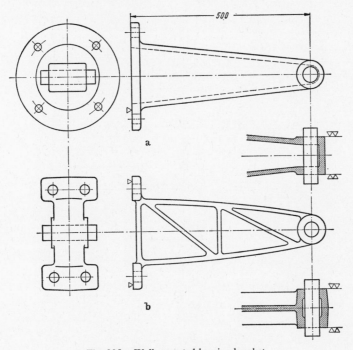

Fig. 285.—Wall-mounted bearing bracket

Problem 11 (p. 91)

One immediately gains the impression that the shape of the casting is not sufficiently compact. Instead, it has a rambling appearance. For this reason one might be tempted to " round off " the shape by using a rib to join the two bearing pedestals (fig. 286*a*). Incidentally, the base would of course be provided with bearer pads. Another idea which suggests itself as a means of avoiding a rambling shape is to separate the pedestals from the base as shown in type *b*. Neither of these designs, however, is satisfactory from the point of view of torsional loading. They are not in themselves torsionally stiff, least of all type *b*, and would not become so until bolted down to the surface receiving them. Torsional stiffness is imparted to the bracket by using a diagonally stiffened base (type *c*). The closed shape shown at *d*, however, is both simpler and more pleasing in appearance.

Problem 12 (p. 91)

The first step, as always, is to ascertain the forces involved. In this instance we are interested in the bearing forces at A and B (fig. 67). These stress the arms in bending and torsion.

The T and I forms (fig. 287) are the most suitable ones to use if the bracket is made as an ordinary ribbed casting.

Hollow cast construction is admittedly very pleasing in appearance, but is expensive owing to the use of cores and to the difficulty of supporting them. For light loading

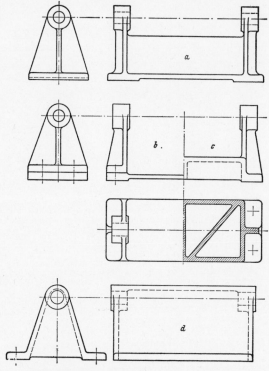

Fig. 286.—Various forms of a bearing bracket

type *a* should be adequate. If larger forces are involved, type *b* will be chosen, whilst type *c* is more costly owing to the greater complexity of the moulding work entailed and offers no advantage over types *a* and *b*.

Problem 13 (p. 92)

In this problem, as in all problems of form design, we will again start by asking about the external and internal forces. The motor transmits a torque of 4 m-kg. The same torque acts on the motor frame to rotate it, and on the pump casing to rotate it likewise but in the opposite direction. The motor and the pump are a fixture with the bedplate, so that the latter must absorb the torque (fig. 288). This factor provides a clue to the way the bedplate must be designed to give adequate strength. The necessary strength is in fact obtained by the provision of diagonal ribbing between the points of application of the forces acting torsionally. Re-checking for torsion might well prove difficult because the formulae available to the designer in handbooks are concerned only with the straightforward cross-sectional forms.

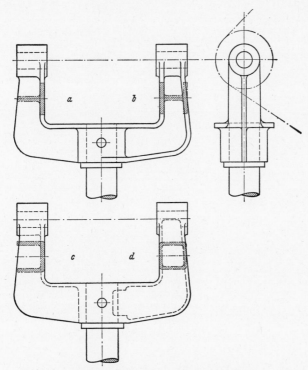

Fig. 287.—Versions of a forked bracket

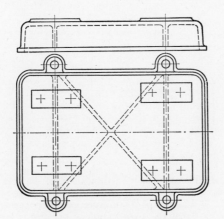

Fig. 288.—Torsionally stiff bedplate

Problem 14 (p. 93)

This casting (fig. 69) is so unfavourably designed that the moulding work entailed, although capable of being carried out, would involve considerable difficulty. We know already that cores must be properly supported and that nipped-in shapes, attachments, and rambling designs are to be avoided. In addition, the core must be capable of easy

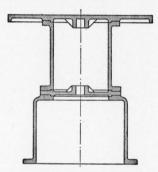

Fig. 289.—Table in cast iron

removal after casting. The pattern in question offends against all of these principles, and for this reason modifications are made. Subdivision of the casting (fig. 289) eliminates all of these disadvantages. The new design will certainly need more machining, but this is not likely to turn out more expensive than the complicated machining work required by the large casting shown in fig. 69.

Problem 15 (p. 93)

The twin lever can be moulded in either of two ways (fig. 290*a* and *b*). Type *a* requires two boxes and a core whilst type *b* requires only two boxes. Fig. 290*b* shows the corresponding form of the arm cross-sections.

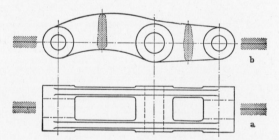

Fig. 290.—Methods of moulding a twin lever

Problem 16 (p. 99)

To begin with, the dimensions of the journals and of the bore and bush can be retained, since there is no undue difference in the strength of forged steel and cast steel. The forging is a typical example of poor utilization of material. For reasons of manufacture, the form is such that the cross-section in the arm is very lightly stressed, whereas in the vulnerable cross-sections the metal is stressed right up to the limit permitted. In this instance, therefore, there is no question of a body of uniform strength.

The situation is different when the item is designed for casting in steel. Under these conditions the designer can choose right from the start the most satisfactory cross-section to withstand the stressing applied. In addition, it is advantageous, in contrast

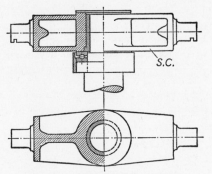

Fig. 291.—Cross head in cast steel

with the forged-steel version, to make the compressive force act from below, so that the whole cross-section can be utilized for carrying the load. The final form of the design for cast steel is shown in fig. 291.

Problem 17 (p. 99)

The eyes at A and B should be bushed with bronze, since bearing loads for cast iron are very low (up to 250lb/in^2) and the bearings will otherwise have to be made unduly large.

Bores A and B can then be given the same size for the cast-steel version. If the forces are small, the cast-iron version can be given arm cross-sections like fig. 292a. Larger forces, as encountered in our example, will require the T-shaped cross-section b which places more metal on the side in tension.

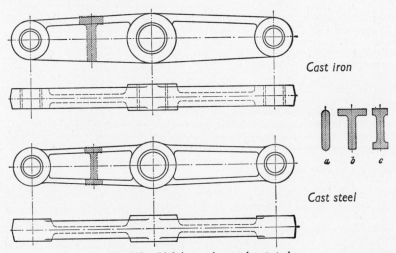

Fig. 292.—Link in cast iron and cast steel

In the design of the cast-steel version there is no need to pay special attention to the tensile forces compared with the compressive forces, as there is with cast iron, and instead either cross-section *a* or *c* can be chosen. The cast-steel link is naturally thinner-walled and is lighter.

Problem 18 (p. 112)

The strength of the welded version is given approximately by its dimensions. If Silumin is used for the cover the strength obtained will not be far different from that of the welded design. To increase the strength a wall thickness of 5 mm is assumed for

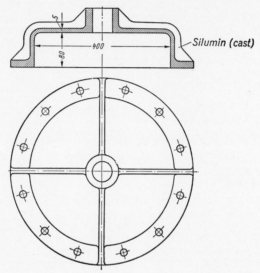

Fig. 293.—Cover

reasons of casting technique. The casting is designed for maximum uniformity of thickness throughout. Increased strength and rigidity are obtained by ribbing the boss, the cover itself and the flange. In view of the lower compression strength of the aluminium alloy, the flange should be made somewhat wider. The elasticity calls for a larger number of bolts. These factors result in the design shown in fig. 293.

Problem 19 (p. 147)

We know already that welded fabrications can be made either in plate only or in a combination of plate and rolled sections.

Since the bracket is bushed in the cast-iron version, it is possible to assume the same dimensions for the welded version. Then, if the walls are made half the thickness of the grey-iron version, we have the four designs illustrated in fig. 294.

The choice as to which kind is the most suitable will, of course, be decided here also by the requirements to be satisfied.

Problem 20 (p. 147)

The solution of this problem is shown in fig. 295*a*, *b*, and *c*. An important point to note in *a* and *c* is the way assembly is simplified by the guidance provided for the pin. Type *b* would probably be the cheapest to make.

The student designer is strongly recommended to vary the size factors in this problem and then to find the appropriate solution.

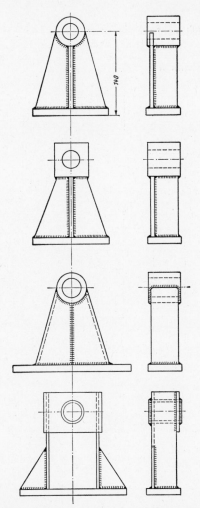

Fig. 294.—Various designs for a steel bearing bracket

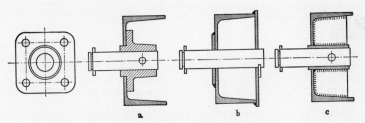

a   b   c

Fig. 295.—Possible solutions to problem 20

248

Problem 21 (p. 147)

Brackets of this kind are of course available commercially. It may happen, however, that an unusual size is needed and this may compel on-site manufacture. In a situation of this kind where only quite a small quantity is required the obvious solution is to use welding on the lines shown in fig. 296.

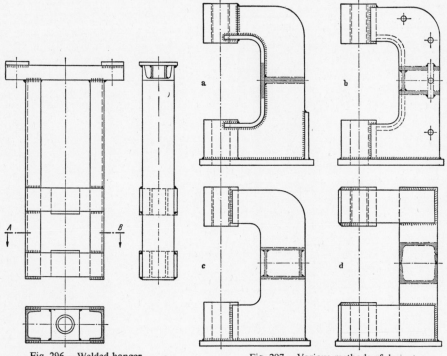

Fig. 296.—Welded hanger.
Section A-B

Fig. 297.—Various methods of designing
the frame of a gap press

Problem 22 (p. 147)

As explained on page 144, the design can be carried out in plate only, or in a combination of rolled sections and plate. Since the unit concerned is comparatively small, the only possibilities are plate construction as in *a* and *b*, hollow construction as in *c*, and fabrication from steel sections as in *d* (fig. 297).

Problem 23 (p. 147)

The designer must see to it that the object he is designing can withstand the external forces acting on it in the optimum manner, that is to say with minimum use of material. The bedplate is acted upon by equal and opposite torques. It should therefore be designed to be torsionally stiff. This condition is obtained in the manner shown in fig. 298.

For fabrication by welding the necessary torsional stiffness can be imparted in various ways. In fig. 298*a* and *b* diagonal ribbing is used. However, a tubular cross-section offers the largest polar moment of inertia for the same amount of metal, and for this reason it can be used with advantage, particularly when welding is employed for fabricating the bedplate (fig. 298c and d).

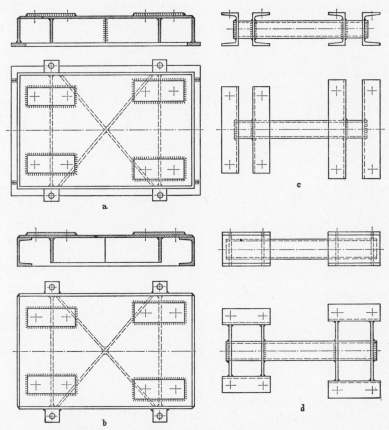

Fig. 298.—Various methods of fabricating a bedplate by welding

Problem 24 (p. 154)

In the early days of mechanical engineering little importance was attached to rational production. Hand forging, for example, can produce the most intricate forms when an expert wields the hammer. One has only to think of the artistic screens, dating from the baroque or rococo periods, wreathed in an abundance of the liveliest flower forms and still capable of exciting astonishment and admiration at the present day.

Modern mechanical engineering, however, demands shapes of the utmost possible simplicity which can be made with ordinary hand tools without a lot of extra appliances being needed.

As far as the crank shown in fig. 180 is concerned, the angled bosses are not easy to make and cannot be given a clean finish without drop forging. The forging operation is greatly simplified by adopting the crank design shown in fig. 299a.

With the forked lever the situation is much the same. Here, too, the round bosses

9

would be better set down square. The forging of the centre boss to a circular shape would be a very difficult and tedious operation. Fig. 299*b* shows the best way to ensure simplicity of manufacture.

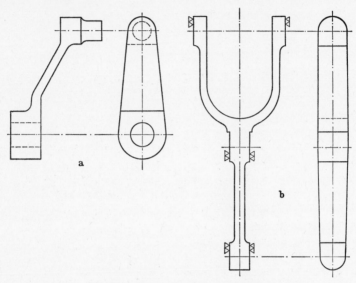

Fig. 299.—Correct form design of forgings

Problem 25 (p. 158)

Proper forging of the tapered surfaces of the crosshead calls for appropriate dies which would have to be made beforehand. The best way therefore is for the designer to avoid tapered surfaces and to design the crosshead as shown in fig. 300 by applying Rule No. 7 (p.157).

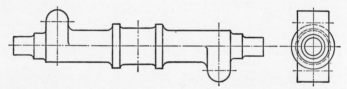

Fig. 300.—Correct form design of a forging

Problem 26 (p. 159)

Fabrication of the frame without resorting to welding is unthinkable. As originally made, the frame was manufactured in the following sequence of operations (fig. 301).

1. Start with a bar equal in thickness to the thickest part of the frame.
2. Notch the bar and draw out the left-hand portion.
3. Punch.
4. Slit.
5. Bend back and work out the split ends.
6. Bend over to form the sides.
7. Weld the two halves of the frame together by hand.
8. Forge the round rods and the eye.

The form of this type of rod will immediately suggest to the present-day designer that welding offers the easiest method of manufacture without requiring much modification to the shape (fig. 302).

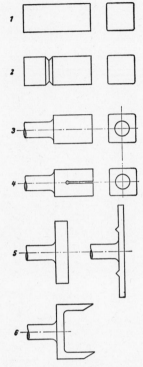

Fig. 301.—Sequence of operations (according to Preger)

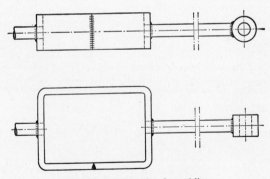

Fig. 302.—Fabrication by welding

Problem 27 (p. 184)

It is impossible to arrange the counterweight in the usual manner, as indicated in fig. 227, and for this reason it must be so mounted that it can swing out without hindrance. The simplest arrangement for achieving this is shown in fig. 303.

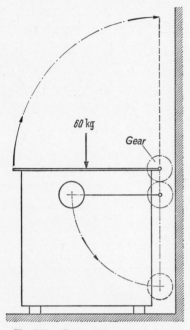

Fig. 303.—Cover with counterweight

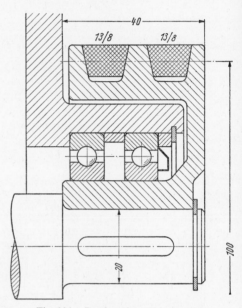

Fig. 304.—Bearings for a V-belt pulley

Problem 28 (p. 184)

Mounting in the prescribed space is only possible if the rolling bearings are positioned between the pulley hub and the housing wall (fig. 304). If the bearings were mounted on the shaft stub it would be necessary to extend the shaft to take the pulley hub, and this is not permissible.

Problem 29 (p. 200)

If the frame is made of two girders the basis of calculation is simplified in contrast with the type shown in fig. 247. The resulting design for welded fabrication is shown in fig. 305. It is clear that a form of this kind saves a good deal of weight compared with the cast-steel version, owing to the fact that the stress conditions in the welded unit can be ascertained much more clearly and are more readily amenable to calculation.

Problem 30 (p. 204)

The riveted frame shown in fig. 250 is already so simple in design (no gusset plates, no stiffening brackets, no laps) that one is inclined to believe that there is no further scope for any substantial simplification coupled with a saving of weight.

The riveted frame will certainly be expected to exhibit a certain amount of torsional stiffness. The most satisfactory way of providing this, in so far as utilization of material

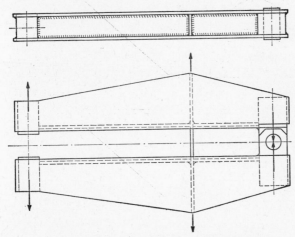

Fig. 305.—Frame built on the lightweight principle

is concerned, is by means of a tube. As a result, the design shown in fig. 306 is torsionally stiffer and at the same time, as shown by calculation, it is 58% lighter than the type shown in fig. 250.

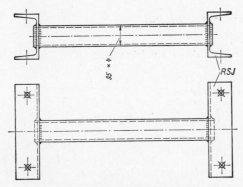

Fig. 306.—Bedplate built on the lightweight principle
Weight = 32.5 kg

Problem 31 (p. 206)

The conversion of a cast-iron bracket to welding always results in a reduction of weight, even when a straight copy of the casting is made, since the higher strength offered by steel enables the wall thicknesses to be reduced to one-half at least. An example of this is given in fig. 307a. By ignoring this example and designing the bracket as shown in fig. 307b, it is possible to cut the weight still further. The lightest weight of all is obtained by welding the item from flat steel bar (fig. 307c).

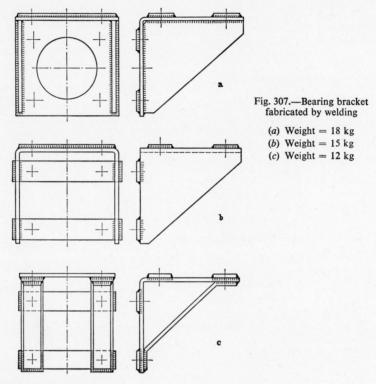

Fig. 307.—Bearing bracket
fabricated by welding

(a) Weight = 18 kg
(b) Weight = 15 kg
(c) Weight = 12 kg

Problem 32 (p. 209)

A rough calculation shows a weight saving of 55% for machinery steel and of 72% for the case-hardened steel (fig. 308), quite apart from the fact that the higher-grade material also results in smaller dimensions.

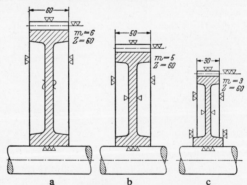

Fig. 308.—How the use of high-grade steel influences size and weight
(a) Cast steel. Weight = 27 kg
(b) Steel. Weight = 12 kg
(c) Case-hardened steel. Weight = 7·5 kg

APPENDIX B

Bibliography

ABBOTT, *Machine Drawing and Design* (Blackie, 1962).

SPANTON, *Geometrical Drawing and Design* (St. Martins).

BUHL, *Creative Engineering Design* (Iowa State, 1960).

GOSLING, *The Design of Engineering Systems* (Heywood, 1962).

GREENWOOD, *Engineering Data for Product Design* (McGraw-Hill).

GREENWOOD, *Product Engineering Design Manual* (McGraw-Hill)

MAREK, *Fundamentals in the Production and Design of Castings* (Wiley)

NIEBEL AND BALDWIN, *Designing for Production* (Irwin, 1957)

FAIRES, *Problems on the Design of Machine Elements* (Macmillan, 1955)

FAIRES and KEOWN, *Mechanism* (McGraw-Hill, 1960).

HALL, *Kinematics and Linkage Design* (Prentice-Hall, 1961)

HINKLE, *Kinematics of Machines* (Prentice-Hall, 1960).

HUNT, *Mechanisms and Motion* (English Universities Press, 1959).

JOHNSON, *Optimum Design of Mechanical Elements* (Wiley, 1961).

KEPLER, *Basical Graphical Kinematics* (McGraw-Hill, 1960).

WATTS, *Machine Design Formulae* (Watts, 1961).

*   *   *   *   *   *   *   *   *   *   *

ANGUS, *Physical and Engineering Properties of Cast Iron* (The British Cast Iron Research Association, 1960).

BARTON, *The Diecasting Process* (Odhams, 1956).

BARTON, *Diecasting Die Design* (Machinery Publishing Co., 1955).

BARTON, *Steel Castings Handbook* (Steel Founders Society of America, 1960).

British and Foreign Specifications for Steel Castings (British Steel Castings Research Association, 1961).

CAMPBELL, *Principles of Manufacturing Materials and Processes* (McGraw-Hill, 1961).

COOK, *Engineered Castings: How to Use, Make, Design and Buy Them* (McGraw-Hill, 1961).

Design for Metal Finishing (Institute of Metal Finishing, 1960).

EARY and REED, *Techniques of Pressworking Sheet Metal: An Engineering Approach to Die Design* (Staples, 1960)

FEES, *Practical Design of Sheet Metal Stampings* (Chapman & Hall, 1960).

FELDMAN, *Cold Forging of Steel* (Hutchinson, 1961).

HERB, *Die Casting* (Machinery Publishing Co., 1952).

JONES, *Fundamental Principles of Powder Metallurgy* (Edward Arnold, 1960).

256

LAING and ROLFE, *A Manual of Foundry Practice for Cast Iron* (Chapman & Hall, 1960).

OATES, *Modern Arc Welding Practice* (Newnes, 1961).

OATES, *Welding Engineers Handbook* (Newnes, 1961).

OATES, *The Caxton Welding Handbook* (Caxton, 1960).

PEARSON and PARKINS, *The Extrusion of Metals* (Chapman & Hall, 1960).

PHILLIPS, *Welding Handbook* (American Welding Society: Cleaver-Hume Press, 1960).

ROLFE, *Malleable Iron Castings* (Malleable Founders Society, Cleveland, Ohio, 1960).

RUSSINOFF, *Forging* (American Technical Society: The Technical Press Ltd., 1952).

TAYLOR, FLEMINGS, WULFF, *Foundry Engineering* (Wiley, 1959).

* * * * * * * * * * *

BEBB, *Plastics Mould Design* (Iliffe, 1962).

British Plastics Year Book (Iliffe).

DELMONTE, *Metal-filled Plastics* (Chapman & Hall, 1962).

MORGAN, *Glass Reinforced Plastics* (Iliffe, 1961).

Plastics International: An Industrial Guide and Catalogue of Plant Materials and Processes (Temple Press, 1961).

Plastics Engineering Handbook (Society of the Plastics Industry, Chapman & Hall, 1960).

APPENDIX C

British Standard Specifications dealing with materials

Steels.

B.S.970 cover a wide range of carbon and alloy steels. There is also an S series covering other special steels.

Grey Iron Castings. B.S.1452.

Aluminium and Aluminium Alloys.

B.S.1476 covers wrought alloys and B.S.1490 cast alloys. B.S.1470–75 deal with particular products as follows:

B.S.1470 Sheet and strip
1471 Drawn tube
1472 Forgings
1473 Rivet, bolt and screw stock
1474 Extruded tube
1475 Wire

There is also an L series covering other aluminium alloys.

Brass (60/40) B.S.1949.

Cast brass (for forging) B.S.944
High-tensile brass B.S.1001–2
Plates, sheet and strip B.S.409

Bronze (phospor)

> Bars and rods B.S.369
> Sheets, strip and foil B.S.407

Aluminimum Bronze B.S.2032

Copper plates for general purposes B.S.2027
 Rolled sheet, strip and foil B.S.899

Zinc (for die casting) B.S.1004

The above list is a small selection only. Full details will be found in the *British Standards Year Book*.

APPENDIX D

Variation of strength of cast iron with wall thickness. Data taken from B.S.1452.

Grade	Wall thickness in inches			
	0·5	1·0	2·0	4·0
10	11·5 / 9·5	8·2 / 7·8	7·8 / 6·5	6·3 / —
12	13·5 / 12·0	10·3 / 9·0	9·7 / 8·2	8·1 / 6·8
14	16·3 / 14·3	13·3 / 11·8	11·8 / 10·2	10·2 / 9·0
17	19·0 / 17·0	15·8 / 14·2	14·3 / 13·0	13·3 / 12·0
20	22·0 / 20·0	19·0 / 17·5	17·0 / 16·0	16·0 / 14·8

The two figures define the range within which the strength lies.

INDEX